APPLIED PROBABILITY

A Series of the Applied Probability Trust

Editors
J. Gani   C.C. Heyde

Gerald S. Shedler

Regeneration and Networks of Queues

With 15 Illustrations

Springer-Verlag
New York Berlin Heidelberg
London Paris Tokyo

Gerald S. Shedler
IBM Almaden Research Center
San Jose, CA 95120-6099
U.S.A.

Series Editors

J. Gani
Statistics Program
Department of Mathematics
University of California
Santa Barbara, CA 93106
U.S.A.

C.C. Heyde
Department of Statistics
Institute of Advanced Studies
The Australian National University
GPO Box 4, Canberra ACT 2601
Australia

AMS Classifications: 60K25, 68C15, 90B22

Library of Congress Cataloging in Publication Data
Shedler, G. S. (Gerald S.), 1933–
  Regeneration and networks of queues.
  (Applied probability)
  Bibliography: p.
  Includes indexes.
  1. Queuing theory. I. Title. II. Series.
T57.9.S53   1987   519.8'2   86-24806

© 1987 Springer-Verlag New York Inc.
All rights reserved. No part of this book may be translated or reproduced in any form without written permission from Springer-Verlag, 175 Fifth Avenue, New York, New York 10010, U.S.A.

Printed and bound by R.R. Donnelley & Sons, Harrisonburg, Virginia.
Printed in the United States of America.

9 8 7 6 5 4 3 2 1

ISBN 0-387-96425-8 Springer-Verlag New York Berlin Heidelberg
ISBN 3-540-96425-8 Springer-Verlag Berlin Heidelberg New York

Preface

Networks of queues arise frequently as models for a wide variety of congestion phenomena. Discrete event simulation is often the only available means for studying the behavior of complex networks and many such simulations are non-Markovian in the sense that the underlying stochastic process cannot be represented as a continuous time Markov chain with countable state space. Based on representation of the underlying stochastic process of the simulation as a generalized semi-Markov process, this book develops probabilistic and statistical methods for discrete event simulation of networks of queues. The emphasis is on the use of underlying regenerative stochastic process structure for the design of simulation experiments and the analysis of simulation output.

The most obvious methodological advantage of simulation is that in principle it is applicable to stochastic systems of arbitrary complexity. In practice, however, it is often a decidedly nontrivial matter to obtain from a simulation information that is both useful and accurate, and to obtain it in an efficient manner. These difficulties arise primarily from the inherent variability in a stochastic system, and it is necessary to seek theoretically sound and computationally efficient methods for carrying out the simulation. Apart from implementation considerations, important concerns for simulation relate to efficient methods for generating sample paths of the underlying stochastic process, the design of simulation experiments, and the analysis of simulation output. It is fundamental for simulation, since results are based on observation of a stochastic system, that some assessment of the precision of results be provided. Assessing the precision of a point estimate requires careful analysis of the simulation output. In general, the desired statistical precision takes the form of a confidence interval. The regenerative method, based on limit theorems for regenerative stochastic processes, is central to the discussion.

The presentation is self-contained. Some knowledge of elementary probability theory, statistics, and stochastic models is sufficient to understand the estimation procedures and the examples. The derivations use results often contained in a first year graduate course on stochastic processes. A brief review of the necessary material is in Appendix 1.

I am indebted to Donald Iglehart for his encouragement and have benefitted from the comments of Peter Haas and my students in the Department of Operations

Research at Stanford University. I am grateful to the IBM Corporation for support of my work and for providing resources of the Almaden Research Center indispensable for the writing of this book.

Los Gatos, California
June, 1986   Gerald S. Shedler

Contents

	Preface	v
1	*Discrete Event Simulation*	1
1.1	Methodological Considerations	1
1.2	The Generalized Semi-Markov Process Model	2
1.3	Specification of Discrete Event Simulations	7
2	*Regenerative Simulation*	20
2.1	Regenerative Stochastic Processes	20
2.2	Properties of Regenerative Processes	24
2.3	The Regenerative Method for Simulation Analysis	27
2.4	Implementation Considerations	34
2.5	Theoretical Values for Discrete Time Markov Chains	37
2.6	Theoretical Values for Continuous Time Markov Chains	41
2.7	Efficiency of Regenerative Simulation	46
2.8	Regenerative Generalized Semi-Markov Processes	50
3	*Markovian Networks*	58
3.1.	Markovian Job Stack Processes	58
3.2.	Augmented Job Stack Processes	76
3.3.	Irreducible, Closed Sets of Recurrent States	80
3.4.	The Marked Job Method	86
3.5.	Fully Augmented Job Stack Processes	97
3.6.	The Labelled Jobs Method	100
3.7.	Sequences of Passage Times	103
3.8.	Networks with Multiple Job Types	110
3.9.	Simulation for Passage Times	116
4	*Non-Markovian Networks*	137
4.1	Networks with Single States	137
4.2	Regenerative Simulation of Non-Markovian Networks	141
4.3	Single States for Passage Times	146

4.4	Recurrence and Regeneration	153
4.5	The Marked Job Method	155
4.6	Finite Capacity Open Networks	164
4.7	Passage Through Subnetworks	173
4.8	The Underlying Stochastic Structure	176
4.9	The Labelled Jobs Method	180
4.10	Comparison of Methods	184

Appendix 1 Limit Theorems for Stochastic Processes	196
Appendix 2 Convergence of Passage Times	210
Bibliography	213
Symbol Index	218
Subject Index	221

Chapter 1

Discrete Event Simulation

It appears to be the rule rather than the exception that usefully detailed stochastic models are sufficiently complex so that it is extremely difficult or impossible to obtain an exact analytic solution. Simulation is essentially a controlled statistical sampling technique that can be used to study complex stochastic systems when analytic techniques do not suffice. This book concentrates on *discrete event digital simulation* in which the behavior of a specified stochastic system is observed by sampling on a digital computer system and stochastic state transitions occur only at a set of increasing (random) time points.

1.1 Methodological Considerations

Implicit in the implementation of any simulation is the definition of an appropriate *system state* that maintains sufficient information so that state transitions that occur over time (together with the times at which these transitions occur) determine the system characteristics of interest. This "state of the system at time t" defines a stochastic process in continuous or discrete time. (Most of the stochastic processes encountered in a discrete event simulation have piecewise constant sample paths.) As it evolves in time, the behavior of this underlying stochastic process is observed when carrying out the simulation. It is necessary to have a means of generating sample paths of these processes. It is also necessary to have a method of obtaining meaningful estimates for the system characteristics of interest. Since point estimates based on small samples can be very misleading, it is important to determine what constitutes an adequate simulation experiment. In particular, we must decide what measurements (observations) to make and how to combine them in order to obtain a *point estimate* for the system characteristics. The usual purpose of a simulation is to provide quantitative information so that a decision can be made among alternatives. This implies that it is necessary to observe the behavior of the system under a number of different circumstances and to draw inferences concerning the resulting differences in system behavior. The simulation analyst must be able to distinguish real differences from apparent differences that are due to random fluctuation.

When reporting simulation results it is absolutely essential to provide some indication of the precision of the estimate. Estimates of greater precision obtainable for equal computational work (cost) are clearly preferable. Assessing the precision of a point estimate requires careful design of the simulation experiments and analysis of the simulation output. In general, the desired precision takes the form of a *confidence interval* for the quantity of interest. Methods are needed for selection of initial conditions for the system, length of the simulation run, number of replications of the experiments, and length of the confidence interval. These problems are difficult primarily because the observations made in the course of a discrete event simulation are generally far from being independent and identically distributed as in classical statistics. A theory of simulation output analysis based on limit theorems for stochastic processes is central to the discussion.

1.2 The Generalized Semi-Markov Process Model

Simulations of networks of queues are often *non-Markovian* in the sense that the underlying stochastic process of the simulation cannot be represented as a Markov chain with countable state space. We focus on simulation methods for non-Markovian

1.2 The Generalized Semi-Markov Model

networks of queues in continuous time and restrict attention to simulations with an underlying stochastic process that can be represented as a *generalized semi-Markov process* (GSMP).

Heuristically, a GSMP is a stochastic process that makes a state transition when an event associated with the occupied state occurs. Each of the several possible events associated with a state competes with respect to triggering the next transition and each of these events has its own distribution for determining the next state. At each transition of the GSMP, new events may be scheduled. For each of these new events, a clock indicating the time until the event is scheduled to occur is set according to an independent (stochastic) mechanism. If a scheduled event does not trigger a transition but is associated with the next state, its clock continues to run; if such an event is not associated with the next state, it ceases to be scheduled and its clock reading is abandoned.

Formal definition of a GSMP is in terms of a *general state space Markov chain* (GSSMC) that describes the process at successive epochs of state transition. Let S be a finite or countable set of *states* and $E = \{e_1, e_2, ..., e_M\}$ be a finite set of *events*. For $s \in S$, $E(s)$ denotes the set of all events that can occur when the GSMP is in state s. When the process is in state s, the occurrence of an event $e \in E(s)$ triggers a transition to a state s'. We denote by $p(s'; s, e)$ the probability that the new state is s' given that event e triggers a transition in state s. For each $s \in S$ and $e \in E(s)$ we assume that $p(\cdot; s, e)$ is a probability mass function. The actual event $e \in E(s)$ that triggers a transition in state s depends on *clocks* associated with the events in $E(s)$ and the *speeds* at which these clocks run. Each clock corresponding to an event $e \in E(s)$ records the remaining time until its associated event triggers a state transition. (The reading on a clock associated with an event $e \notin E(s)$ is zero.) We denote by r_{si} (≥ 0) the deterministic rate at which the clock associated with event e_i runs in state s; for each $s \in S$, $r_{si} = 0$ if $e_i \notin E(s)$. We assume that $r_{si} > 0$ for some $e_i \in E(s)$. (Typically in applications, all speeds r_{si} are equal to one. There are, however, models in which speeds other than unity as well as state-dependent speeds are convenient. For example, zero speeds are needed in queueing systems with service interruptions of the preemptive-resume type.)

For $s \in S$ define the set, $C(s)$, of possible clock readings in state s:

$$(2.1) \quad C(s) = \{(c_1,\ldots,c_M): c_i \geq 0 \text{ and } c_i > 0 \text{ if and only if } e_i \in E(s);$$
$$c_i r_{si}^{-1} \neq c_j r_{sj}^{-1} \text{ for } i \neq j \text{ with } c_i c_j r_{si} r_{sj} > 0\}.$$

The conditions in (2.1) ensure that no two events simultaneously trigger a transition (as defined below). The clock with reading c_i and event e_i are said to be *active* in state s if $e_i \in E(s)$. For $s \in S$ and $c \in C(s)$, let

$$(2.2) \quad t^* = t^*(s,c) = \min_{\{i: e_i \in E(s)\}} \{c_i r_{si}^{-1}\},$$

where $c_i r_{si}^{-1}$ is taken to be $+\infty$ when $r_{si} = 0$. Also set

$$(2.3) \quad c_i^* = c_i^*(s,c) = c_i - t^*(s,c) r_{si}, \quad e_i \in E(s)$$

and

$$(2.4) \quad i^* = i^*(s,c) = i \text{ such that } e_i \in E(s) \text{ and } c_i^*(s,c) = 0.$$

Beginning in state s with clock vector c, $t^*(s,c)$ is the time to the next state transition and $i^*(s,c)$ is the index of the unique triggering event

$$e^* = e^*(s,c) = e_{i^*(s,c)}.$$

At a transition from state s to state s' triggered by event e^*, new clock times are generated for each

$$e' \in N(s'; s, e^*) = E(s') - (E(s) - \{e^*\}).$$

The distribution function of such a new clock time is denoted by $F(\cdot; s', e', s, e^*)$ and we assume that $F(0; s', e', s, e^*) = 0$. For

$$e' = O(s'; s, e^*) = E(s') \cap (E(s) - \{e^*\}),$$

the old clock reading is kept after the transition. For $e' \in (E(s) - \{e^*\}) - E(s')$, event e' ceases to be scheduled after the transition and its clock reading is abandoned. See Figure 1.1.

Next consider a GSSMC $\{(S_n, C_n) : n \geq 0\}$ having state space,

$$\Sigma = \bigcup_{s \in S} (\{s\} \times C(s))$$

and representing the state (S_n) and vector (C_n) of clock readings at successive state transition epochs. ($C_{n,i}$ denotes the ith coordinate of the vector C_n.) The transition kernel of the GSSMC is

$$(2.5) \quad P((s,c), A) = p(s'; s, e^*) \prod_{e_i \in N(s')} F(a_i; s', e_i, s, e^*) \prod_{e_i \in O(s')} 1_{[0, a_i]}(c_i^*),$$

where $N(s') = N(s'; s, e^*)$, $O(s') = O(s'; s, e^*)$, and

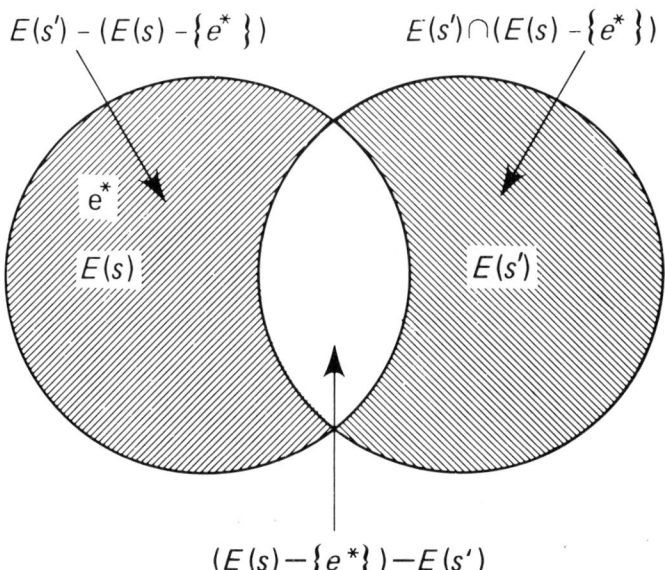

Figure 1.1. Sets of new and old events

$$A = \{s'\} \times \{(c_1',...,c_M') \in C(s'): c_i' \leq a_i \text{ for } e_i \in E(s')\}.$$

The set A is the subset of Σ that corresponds to the GSMP entering state s' with the reading c_i' on the clock associated with event $e_i \in E(s')$ set to a value in $[0, a_i]$. (We assume that $P((s,c), \Sigma) = 1$.)

Finally, the GSMP is a piecewise constant continuous time process constructed from the GSSMC $\{(S_n, C_n) : n \geq 0\}$ in the following manner. Set $\zeta_0 = 0$ and

$$\zeta_n = \sum_{k=0}^{n-1} t^*(S_k, C_k),$$

$n \geq 1$. According to this definition ζ_n is the nth time at which the process makes a state transition. (Throughout we assume that

$$P\{\sup_{n \geq 1} \zeta_n = \infty \mid (S_0, C_0)\} = 1 \text{ a.s.}$$

for all initial states (S_0, C_0).) Then set

(2.6) $$X(t) = S_{N(t)},$$

where

(2.7) $$N(t) = \max\{n \geq 0 : \zeta_n \leq t\}.$$

The process $\{X(t) : t \geq 0\}$ defined by (2.6) is a GSMP.

Proposition 2.10 prescribes conditions under which a finite state GSMP has a limiting distribution in the sense that there exists a random variable X such that

(2.8) $$\lim_{t \to \infty} P\{X(t) \leq x\} = P\{X \leq x\}$$

for all x at which the right hand side of (2.8) is continuous. This type of convergence in distribution is known as *weak convergence* and is denoted by "$X(t) \Rightarrow X$ as $t \to \infty$."

Definition 2.9. A GSMP having state space, S, and event set, E, is said to be *irreducible* if for each pair $s,s' \in S$ there exists a finite sequence of states $s_1, s_2, \ldots, s_n \in S$ and events $e_{i_0}, e_{i_1}, \ldots, e_{i_n} \in E$ such that

$$p(s_1;s,e_{i_0})r_{si_0}p(s_2;s_1,e_{i_1})r_{s_1 i_1} \cdots p(s';s_n,e_{i_n})r_{s_n i_n} > 0.$$

Proposition 2.10. Let $\{X(t): t \geq 0\}$ be an irreducible GSMP with a finite state space, S, and event set, E. Suppose that for all $s, s' \in S$, $e^* \in E$, and $e' \in N(s'; s, e^*)$ the clock setting distribution $F(\cdot; s', e', s, e^*)$ has a finite mean and a density function that is continuous and positive on $(0, +\infty)$. Then $X(t) \Rightarrow X$ as $t \to \infty$.

This proposition provides a means of establishing the existence of a "steady state" for a discrete event simulation model. It is sufficient to show that the underlying stochastic process of the simulation can be represented as an irreducible, finite state GSMP in which all clock setting distributions have finite mean and a density function that is continuous and positive on $(0, +\infty)$.

1.3 Specification of Discrete Event Simulations

In this section, we illustrate the use of the GSMP framework for formal representation of the underlying stochastic process of a queueing system simulation.

Example 3.1. (Cyclic Queues With Feedback) Consider a queueing system consisting of two single-server service centers and a fixed number, N, of jobs; see Figure 1.2. After service completion at center 1, a job moves instantaneously to the tail of the queue at center 1 with fixed probability p $(0 < p < 1)$ and (with probability $1 - p$) moves to the tail of the queue at center 2. After service completion at center 2, a job moves to the tail of the queue in center 1. Assume that both queues are served according to a first-come, first-served (FCFS) discipline. Also suppose that all service times are mutually independent and that the service times at center i are identically distributed as a positive random variable, L_i, $i = 1,2$.

Let $X(t)$ be the number of jobs waiting or in service at center 2 at time t. The process $\{X(t): t \geq 0\}$ is a GSMP with finite state space, $S = \{0,1,...,N\}$, and event set, $E = \{e_1, e_2\}$, where event $e_i = $ "service completion at center i." For $s \in S$ the sets $E(s)$ of events that can occur in state s are as follows. The event $e_1 \in E(s)$ if and only if $0 \leq s < N$ and the event $e_2 \in E(s)$ if and only if $0 < s \leq N$. If $e = e_1$, then the state transition probability $p(s + 1; s, e) = 1 - p$ and $p(s; s, e) = p$ when $0 \leq s < N$; if $e = e_2$ then $p(s - 1; s, e) = 1$ when $0 < s \leq N$. All other state transition probabilities $p(s'; s, e)$ are equal to zero.

The set $O(s - 1; s, e_2)$ of old events (when event e_2 triggers a transition from state s to state $s - 1$) equals $\{e_1\}$ if $0 < s < N$ and equals $\emptyset$ if $s = N$. The set $O(s + 1; s, e_1)$ of old events equals $\{e_2\}$ if $0 < s < N$ and equals $\emptyset$ if $s = 0$. The set $N(s + 1; s, e_1)$ of new

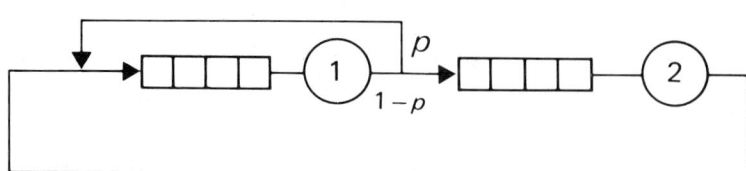

Figure 1.2. Cyclic queues with feedback

events (when event e_1 triggers a transition from state s to state $s + 1$) equals $\{e_1,e_2\}$ if $s = 0$, equals $\{e_1\}$ if $1 \le s < N - 1$, and equals $\emptyset$ if $s = N - 1$. The set $N(s;s,e_1)$ of new events (when event e_1 triggers a transition from state s to state s) equals $\{e_1\}$ if $0 \le s < N$. The set $N(s - 1;s,e_2)$ equals $\{e_1,e_2\}$ if $s = N$, equals $\{e_2\}$ if $1 < s < N$, and equals $\emptyset$ if $s = 1$. The distribution functions of new clock times for events $e' \in N(s';s,e^*)$ are as follows. If $e' = e_i$, then the distribution function $F(x;s',e',s,e^*) = P\{L_i \le x\}$.

Example 3.2. (Two-server Cyclic Queues) Consider a queueing system consisting of a fixed number, N, of jobs and two service centers, each having two servers. After service completion at center 1, a job moves instantaneously from center 1 to the tail of the queue in center 2. After service completion at center 2, a job moves back to the tail of the queue in center 1. If both servers at a center are idle when a job joins the tail of the queue at a center, the job is served by server 1. Suppose that that both queues are served according to a FCFS discipline. Also suppose that all service times are mutually independent and that the service times at center i are identically distributed as a positive random variable, L_i, with finite mean, $i = 1,2$.

For $t \ge 0$ set

(3.3) $\qquad X(t) = (Z_1(t), S_{11}(t), S_{12}(t), S_{21}(t), S_{22}(t))$,

where $Z_1(t)$ is the number of jobs waiting or in service at center 1 at time t and $S_{ij}(t)$ equals one if server j at center i is busy at time t. ($S_{ij}(t)$ equals zero if server j at center i is idle at time t.) Set $S^* = \{0,1,...,N\} \times (\{0,1\})^4$. The process $\{X(t): t \ge 0\}$ is a GSMP with state space,

$$S = \{(z_1, s_{11}, s_{12}, s_{21}, s_{22}) \in S^* : s_{11} + s_{12} = 2 \wedge z_1;$$

$$s_{21} + s_{22} = 2 \wedge (N - z_1)\}$$

and event set, $E = \{e_{11}, e_{12}, e_{21}, e_{22}\}$, where event e_{ij} = "service completion by server j at center i." (Recall that $x \wedge y$ equals x if

1 Discrete Event Simulation

$x \leq y$ and equals y otherwise.) For $s = (z_1, s_{11}, s_{12}, s_{21}, s_{22}) \in S$, the event $e_{ij} \in E(s)$ if and only if $s_{ij} = 1$.

If $e = e_{12}$, then the state transition probability $p(s'; s, e) = 1$ when

$$s = (z_1, s_{11}, 1, 0, s_{22}) \quad \text{and} \quad s' = (z_1 - 1, s_{11}, 0, 1, s_{22})$$

with $0 < z_1 \leq N$, when

$$s = (z_1, s_{11}, 1, 1, 0) \quad \text{and} \quad s' = (z_1 - 1, s_{11}, 0, 1, 1)$$

with $0 < z_1 \leq N$, and when

$$s = (z_1, s_{11}, 1, 1, 1) \quad \text{and} \quad s' = (z_1 - 1, s_{11}, 0, 1, 1)$$

with $0 < z_1 \leq N$. If $e = e_{11}$, then the state transition probability $p(s'; s, e) = 1$ when

$$s = (z_1, 1, s_{12}, 0, s_{22}) \quad \text{and} \quad s' = (z_1 - 1, 0, s_{12}, 1, s_{22})$$

with $0 < z_1 \leq N$, when

$$s = (z_1, 1, s_{12}, 1, 0) \quad \text{and} \quad s' = (z_1 - 1, 0, s_{12}, 1, 1)$$

with $0 < z_1 \leq N$, and when

$$s = (z_1, 1, s_{12}, 1, 1) \quad \text{and} \quad s' = (z_1 - 1, 0, s_{12}, 1, 1)$$

with $0 < z_1 \leq N$. If $e = e_{22}$, then the state transition probability $p(s'; s, e) = 1$ when

$$s = (z_1, 0, s_{12}, s_{21}, 1) \quad \text{and} \quad s' = (z_1 + 1, 1, s_{12}, s_{21}, 0)$$

with $0 \leq z_1 < N$, when

$$s = (z_1, 1, 0, s_{21}, 1) \quad \text{and} \quad s' = (z_1 + 1, 1, 1, s_{21}, 0)$$

with $0 \leq z_1 < N$, and when

$$s = (z_1, 1, 1, s_{21}, 1) \quad \text{and} \quad s' = (z_1 + 1, 1, 1, s_{21}, 0)$$

with $0 \leq z_1 < N$. If $e = e_{21}$, then the state transition probability $p(s'; s, e) = 1$ when

$$s = (z_1, 0, s_{12}, 1, s_{22}) \quad \text{and} \quad s' = (z_1 + 1, 1, s_{12}, 0, s_{22})$$

with $0 \leq z_1 < N$, when

$$s = (z_1, 1, 0, 1, s_{22}) \quad \text{and} \quad s' = (z_1 + 1, 1, 1, 0, s_{22})$$

with $0 \leq z_1 < N$, and when

$$s = (z_1, 1, 1, 1, s_{22}) \quad \text{and} \quad s' = (z_1 + 1, 1, 1, 0, s_{22})$$

with $0 \leq z_1 < N$. All other state transition probabilities $p(s';s,e)$ are equal to zero. For $s, s' \in S$ and $e^* \in E(s)$, the sets $N(s';s,e^*)$ of new events when event e^* triggers a transition from state s to state s' are as follows. Let $s = (z_1, s_{11}, s_{12}, s_{21}, s_{22}) \in S$, and set $z_2 = N - z_1$. For $e^* = e_{i1}$ ($i = 1,2$) and $k = 3 - i$, the event $e_{i1} \in N(s';s,e^*)$ if and only if $z_i > 1 + s_{i2}$; $e_{k1} \in N(s';s,e^*)$ if and only if $s_{k1} = 0$; and $e_{k2} \in N(s';s,e^*)$ if and only if $s_{k1} = 1$ and $s_{k2} = 0$. For $e^* = e_{i2}$ and $k = 3 - i$, the event $e_{i2} \in N(s';s,e^*)$ if and only if $3 \leq z_i \leq N$; $e_{k1} \in N(s';s,e^*)$ if and only if $s_{k1} = 0$; and $e_{k2} \in N(s';s,e^*)$ if and only if $s_{k1} = 1$ and $s_{k2} = 0$. The distribution functions of new clock times for events $e' \in N(s';s,e^*)$ are as follows. If $e' = e_{ij}$, then the distribution function $F(x;s',e',s,e^*) = P\{L_i \leq x\}$, $i = 1,2$.

Example 3.4. (System Overhead Model) Consider a queueing system that consists of two stages, the α-stage and the β-stage, in a loop. (Each stage comprises a queue and three sequential services.) Two servers provide service to N (≥ 2) stochastically identical jobs. Each job goes through the α-stage and the β-stage in sequence and then returns to the α-stage, this process being repeated indefinitely. Within the α-stage, a job receives each of three services, α_1, α_2, and α_3, in that order; similarly, within the β-stage, a job receives each of three services, β_1, β_2, and β_3, in that order. Only server 2 can provide a β_2 service and only server 1 can provide an $\alpha_1, \alpha_2, \alpha_3, \beta_1$, or β_3 service. The two servers can provide service concurrently, subject to the restriction that server 1 cannot provide a β_1 or a β_3 service while server 2 is providing a β_2 service. After having received an α_3 service, a job moves instantaneously from the α-stage to the tail of the queue in the β-stage and after having received a β_3 service, moves instantaneously from the β-stage to the tail of the queue in the α-stage. Interruption of an α_2 service occurs at the completion of a concurrent β_2 service. The "β_2-complete" interruption of an α_2 service is of the preemptive-resume type. All other services in the system complete without interruption.

At the completion of an α_1, α_2, α_3, β_1, or β_3 service or at an interruption of an α_2 service, server 1 chooses the next service according to a rule of priority as follows:

(i) if there is a job waiting for β_3 service, begin this service;

(ii) if there is a job waiting for β_1 service, begin this service unless β_2 service is in progress;

(iii) if the last α-stage service provided was a completed α_2 service, begin an α_3 service;

(iv) if the last α-stage service provided was an interrupted α_2 service, resume the α_2 service;

(v) if the last α-stage service provided was an α_1 service, begin an α_2 service;

(vi) if the last α-stage service provided was an α_3 service and if the queue in the α-stage is not empty, begin an α_1 service.

If no claim is made on server 1 according to the rule of priority, server 1 remains idle until the completion of the next β_2 service, at which time the rule of priority is invoked again. The queue service discipline in the α-stage and in the β-stage is FCFS. (An interpretation of the six services α_1, α_2, α_3, β_1, β_2, and β_3, in a multiprogrammed computer system is as follows. Problem program processing corresponds to α_2 service and data transfer service (paging) corresponds to β_2 service. The remaining services α_1, α_3, β_1, and β_3 are interpreted as system overhead functions.)

Suppose that α_2 service time is an exponentially distributed random variable, A_2, and that α_j service time is a positive random variable, A_j, with finite mean but otherwise arbitrary distribution,

$j = 1,3$. Also suppose that β_j service time is a positive random variable, B_j, with finite mean but otherwise arbitrary distribution, $j = 1,2,3$.

This system can be represented as a closed network of queues with N jobs, $s = 2$ service centers, and $c = 6$ job classes. At every epoch of continuous time each job is of exactly one job class, but jobs may change class as they traverse the network. Upon completion of service at center i a job of class j goes to center k and changes to class l with probability $p_{ij,kl}$, where

$$P = \{p_{ij,kl}:(i,j),(k,l) \in C\}$$

is a given irreducible stochastic matrix and $C = \{(1,1),(1,2),(1,3),(1,4),(1,5),(2,6)\}$ is the set of (center, class) pairs in the network. (Job class 1 corresponds to β_3 service, class 2 to β_1 service, class 3 to α_3 service, class 4 to α_2 service, and class 5 to α_1 service. Job class 6 corresponds to β_2 service.) There is a priority ordering of job classes served at center 1. In order of decreasing priority, the priority ordering is 1, 2, 3, 4, and 5. These priorities are subject to the constraint that there can be no class 1 or class 2 service at center 1 during (class 6) service at center 2.

Service to a job of class 4 at center 1 is subject to preemption when any other job of higher priority joins the queue at center 1. (The interruption of class 4 service is of the preemptive-resume type.) All other services in the network are not interruptable. The job routing matrix P is

$$P = \begin{matrix} 0 & 0 & 0 & 0 & 1 & 0 \\ 0 & 0 & 0 & 0 & 0 & 1 \\ 0 & 1 & 0 & 0 & 0 & 0 \\ 0 & 0 & 1 & 0 & 0 & 0 \\ 0 & 0 & 0 & 1 & 0 & 0 \\ 1 & 0 & 0 & 0 & 0 & 0 \end{matrix}.$$

For $t \geq 0$ set

(3.5) $\qquad X(t) = (C_5(t), C_4(t),...,C_1(t), S_1(t), Q_2(t)),$

where $Q_2(t)$ is the number of jobs waiting or in service at center 2, $S_1(t)$ is the job class in service at center 1 and $C_j(t)$ is the number of class j jobs in queue at center 1 at time t, $j = 1,2,...,5$. ($S_1(t)$ equals 0 if there is no job in service at center 1.) Set

$$S^* = \{0,1,...,N\} \times \{0,1\} \times \{0,1\} \times \{0,1,...,N\} \times \{0,1\} \times \{0,1,...,5\} \times \{0,1\}.$$

The process $\{X(t): t \geq 0\}$ defined by (3.5) is a GSMP with state space,

$$S = \{(c_5,...,c_1,s_1,q_2) \in S^* : c_5 + ... + c_1 + 1_{\{s_1 > 0\}} + q_2 = N;\ c_4 c_3 = 0;$$
$$c_4, c_3 = 0 \text{ when } s_1 \geq 3;\ s_1 \neq 1,2 \text{ and } c_1 = 0 \text{ when } q_2 = 1\}$$

and event set,

$$E = \{e_{11}, e_{12}, e_{13}, e_{14}, e_{15}, e_{26}\},$$

where event e_{ij} = "service completion to a job of class j at center i." For $s = (c_5,...,c_1,s_1,q_2) \in S$, the event sets $E(s)$ are as follows. Event $e_{1j} \in E(s)$ if and only if $s_1 = j$, $j = 1,2,...,5$. Event $e_{26} \in E(s)$ if and only if $q_2 = 1$.

If $e = e_{11}$, then the state transition probability $p(s'; s, e) = 1$ when

$$s = (c_5, c_4, c_3, c_2, 0, 1, 0) \in S \quad \text{and} \quad s' = (c_5 + 1, c_4, c_3, c_2 - 1, 0, 2, 0)$$

with $c_2 > 0$, when

$$s = (c_5, 0, 0, 0, 0, 1, 0) \in S \quad \text{and} \quad s' = (c_5, 0, 0, 0, 0, 5, 0)$$

with $c_5 > 0$, and when

$$s = (c_5, 1, 0, 0, 0, 1, 0) \in S \quad \text{and} \quad s' = (c_5 + 1, 0, 0, 0, 0, 4, 0).$$

If $e = e_{12}$, then $p(s'; s, e) = 1$ when

$$s = (c_5, 1, 0, c_2, 0, 2, 0) \in S \quad \text{and} \quad s' = (c_5, 0, 0, c_2, 0, 4, 1),$$

when
$$s = (0,0,0,c_2,0,2,0) \in S \quad \text{and} \quad s' = (0,0,0,c_2,0,0,1),$$

and when
$$s = (c_5,0,0,c_2,0,2,0) \in S \quad \text{and} \quad s' = (c_5 - 1,0,0,c_2,0,5,1).$$

If $e = e_{13}$, then $p(s';s,e) = 1$ when
$$s = (c_5,0,0,c_2,1,3,0) \in S \quad \text{and} \quad s' = (c_5,0,0,c_2 + 1,0,1,0),$$

when
$$s = (N - 1,0,0,0,0,3,0) \in S \quad \text{and} \quad s' = (N - 1,0,0,0,0,2,0),$$

when
$$s = (c_5,0,0,c_2,0,3,0) \in S \quad \text{and} \quad s' = (c_5,0,0,c_2,0,2,0)$$

with $c_2 > 0$, and when
$$s = (c_5,0,0,c_2,0,3,1) \in S \quad \text{and} \quad s' = (c_5 - 1,0,0,c_2,0,5,1).$$

with $c_5 > 0$. If $e = e_{14}$, then $p(s';s,e) = 1$ when
$$s = (c_5,0,0,c_2,0,4,q_2) \in S \quad \text{and} \quad s' = (c_5,0,0,c_2,0,3,q_2).$$

If $e = e_{15}$, then $p(s';s,e) = 1$ when
$$s = (c_5,0,0,c_2,0,5,1) \in S \quad \text{and} \quad s' = (c_5,0,0,c_2,0,4,1),$$

and when
$$s = (c_5,0,0,c_2,1,5,0) \in S \quad \text{and} \quad s' = (c_5,1,0,c_2,0,1,0).$$

If $e = e_{26}$, then $p(s';s,e) = 1$ when
$$s = (c_5,c_4,c_3,c_2,0,0,1) \in S \quad \text{and} \quad s' = (c_5,c_4,c_3,c_2,0,1,0),$$

when
$$s = (c_5,0,0,c_2,0,3,1) \in S \quad \text{and} \quad s' = (c_5,0,0,c_2,1,3,0),$$

when
$$s = (c_5,0,0,c_2,0,4,1) \in S \quad \text{and} \quad s' = (c_5,1,0,c_2,0,1,0),$$

and when

$$s = (c_5,0,0,c_2,0,5,1) \in S \quad \text{and} \quad s' = (c_5,0,0,c_2,1,5,0).$$

All other state transition probabilities $p(s';s,e)$ are equal to zero.

The distribution functions $F(\cdot;s',e',s,e^*)$ of new clock times for events $e' \in N(s';s,e^*)$ are as follows. If $e' = e_{1j}$ and $s' = (c_5,...,c_1,j,q_2)$ with $j = 3$, 4, or 5, then $F(x;s',e',s,e^*) = P\{A_{6-j} \le x\}$ for all s and e^* such that $p(s';s,e^*) > 0$. If $e' = e_{1j}$ and $s' = (c_5,...,c_1,j,q_2)$ with $j = 1$ or 2, then $F(x;s',e',s,e^*) = P\{B_{5-2j} \le x\}$. If $e' = e_{26}$ and $s' = (c_5,...,c_1,s_1,1)$, then $F(x;s',e',s,e^*) = P\{B_2 \le x\}$. Note that when event $e^* = e_{26}$ triggers a transition from a state $s = (c_5,0,0,c_2,4,1) \in S$ to state $s' = (c_5,1,0,c_2,0,1,0)$, $e_{14} \in (E(s) - \{e^*\}) - E(s')$ so that event e_{14} that was scheduled in state s ceases to be scheduled.

Example 3.6 illustrates the use of zero speeds in a GSMP.

Example 3.6. (Preemptive Cyclic Queues) Consider a queueing system consisting of two single-server service centers and a fixed number, N, of jobs labelled $1,2,...,N$. After service completion at center 1, a job moves instantaneously from center 1 to the tail of the queue in center 2, and after service at center 2 moves back to the tail of the queue in center 1. Suppose that both queues are served according to a last-come, first-served (LCFS) preemptive-resume discipline. (Jobs waiting in queue appear in the order of preemption at the center, the most recently preempted job at the head of the queue.) Also suppose that all service times are mutually independent and that the service times at center i are identically distributed as a positive random variable, $i = 1,2$.

Let $Z_1(t)$ be the number of jobs waiting or in service at center 1 at time t. Define the position, $N_i(t)$, of job i at time t as follows: $N_i(t) = j$ if job i is waiting in queue at center 1 at time t and there are $j - 1$ jobs waiting at center 1 behind job i, $j = 1,2,...,N$; $N_i(t) = Z_1(t)$ if job i is in service at center 1; $N_i(t) = Z_1(t) + j$ if job i is waiting in queue at center 2 and there are $j - 1$ jobs waiting at center 2 behind job i; and $N_i(t) = N$ if job i is in service at center 2. Set

$$X(t) = (Z_1(t), N_1(t), N_2(t), \ldots, N_N(t)).$$

Set $S^* = \{0, 1, \ldots, N\} \times (\{1, 2, \ldots, N\})^N$. The process $\{X(t): t \geq 0\}$ is a GSMP with state space,

$$S = \{(z_1, n_1, \ldots, n_N) \in S^* : n_i \neq n_j \text{ for all } i, j = 1, 2, \ldots, N \text{ with } i \neq j\}$$

and event set

$$E = \{e_1, e_2, \ldots, e_N\},$$

where event $e_i =$ "service completion to job i." The event set $E(s) = E$ for all $s \in S$. The speed, r_{si}, at which the clock associated with event e_i runs in state $s = (z_1, n_1, \ldots, n_N)$ is equal to one when $n_i = z_1$ and when $n_i = N$. All other speeds r_{si} are equal to zero. For $0 < z_1 \leq N$, the state transition probability $p(s'; s, e_i) = 1$ when

$$s = (z_1, n_1, \ldots, n_{i-1}, z_1, n_{i+1}, \ldots, n_N)$$

and

$$s' = (z_1 - 1, n_1', \ldots, n_{i-1}', N, n_{i+1}', \ldots, n_N'),$$

where n_j' equals n_j if $n_j < z_1$ and equals $n_j - 1$ if $n_j > z_1$, $i \neq j$. For $0 \leq z_1 < N$, the state transition probability $p(s'; s, e_i) = 1$ when

$$s = (z_1, n_1, \ldots, n_{i-1}, N, n_{i+1}, \ldots, n_N)$$

and

$$s' = (z_1 + 1, n_1', \ldots, n_{i-1}', z_1 + 1, n_{i+1}', \ldots, n_N'),$$

where n_j' equals n_j if $n_j < z_1$ and equals $n_j + 1$ if $n_j > z_1$, $i \neq j$. All other state transition probabilities $p(s'; s, e)$ are equal to zero.

For many discrete event simulations, a "natural" state definition that maintains information sufficient to determine the system characteristics of interest leads to a stochastic process that is a GSMP. This GSMP representation provides an algorithm for generating sample paths of the process. There are systems, however, for which a natural state definition leads to a stochastic process that is not a GSMP. Often this is because the system state alone does not determine the set of events that are scheduled; i.e., there is no mapping that takes a state s into an event set, $E(s)$, the set of all events scheduled when the system is in state s. Another source of difficulty is that the sets of new events and old events when event e^* triggers a transition from state s to state s' can depend (not only on e^*, s, and s' but also) on c^*, the vector of clock readings when event e^* occurs. When (in addition to e^*, s, and s') the sets of new events and old events depend only on which clock readings in c^* are positive, direct augmentation of the state vector leads to a stochastic process that is a GSMP. The simplest example appears to be a multi-server queue.

Example 3.7. (Multi-server Queue) Consider a queue with K (≥ 2) servers, numbered $1,2,...,K$. Suppose that customers in the queue are served on a first-come, first-served (FCFS) basis, that no server is idle when there is a customer in the queue, and that the waiting room has infinite capacity. If more than one server is idle when a customer arrives, the customer is served by the lowest numbered idle server. Also suppose that the interarrival times and service times, respectively, are independent and identically distributed positive random variables.

Let $X(t)$ be the number of customers waiting or in service at time t. The stochastic process $\{X(t): t \geq 0\}$ has countable state space, $S = \{0,1,...,\}$, but there is no finite set, E, such that the process is a GSMP with state space, S, and event set, E. To illustrate, set

(3.8) $$E = \{e_0, e_1, ..., e_K\},$$

where event $e_0 = $ "arrival of customer" and event $e_i = $ "service completion by server i," $i = 1, 2, ..., K$. The process $\{X(t): t \geq 0\}$ is

1.3 Specification of Discrete Event Simulations

not a GSMP because it cannot be determined (from s alone) whether or not any one of the events $e_1, e_2, \ldots, e_K$ is scheduled when the process is in state s, $0 < s < K$. (When the process is in state $s = 0$, only event e_0 is scheduled: $E(0) = \{e_0\}$; each of the events $e_0, e_1, \ldots, e_K$ is scheduled when the process is in state $s \geq K$: $E(s) = E$.)

The sets of new events and old events when event e^* triggers a transition from state s to state s' depend explicitly on (and only on) the set of clocks with positive readings in c^*. Set

$$(3.9) \qquad U(t) = (U_1(t), U_2(t), \ldots, U_K(t)),$$

where $U_i(t)$ equals one if server i is busy at time t (and equals zero otherwise.) Also set $S^* = \{0, 1, \ldots\} \times (\{0,1\})^K$. The process $\{(X(t), U(t)) : t \geq 0\}$ is a GSMP with event set, E, and state space,

$$(3.10) \qquad S = \{(s, u_1, \ldots, u_K) \in S^* : u_1 + \ldots + u_K = K \wedge s\}.$$

Chapter 2

Regenerative Simulation

Heuristically, a regenerative stochastic process has the characteristic property that there exists a sequence of random time points, referred to as regeneration points or regeneration times, at which the process probabilistically restarts. Typically, the times at which a regenerative process probabilistically starts afresh occur when the process returns to some fixed state. The essence of regeneration is that the evolution of the process between any two successive regeneration points is a probabilistic replica of the process between any other two successive regeneration points. In the presence of certain mild regularity conditions, the regenerative structure guarantees the existence of a limiting distribution ("steady state") for the process provided that the expected time between regeneration points is finite. Moreover, the limiting distribution of a regenerative process is determined (as a ratio of expected values) by the behavior of the process between any pair of successive regeneration points. These results have important implications (discussed in Section 2.3) for the analysis of simulation output.

2.1 Regenerative Stochastic Processes

Formal definition of a regenerative process is in terms of "stopping times" for a stochastic process.

Definition 1.1. A *stopping time* for a stochastic process $\{X(t): t \geq 0\}$ is a random variable T (taking values in $[0, +\infty)$) such that for every finite $t \geq 0$, the occurrence or non-occurrence of the event

2.1 Regenerative Stochastic Processes

$\{T \le t\}$ can be determined from the history $\{X(u):u \le t\}$ of the process up to time t.

Definition 1.2. The real (possibly vector-valued) stochastic process $\{X(t):t \ge 0\}$ is a *regenerative process in continuous time* provided that:

(i) there exists a sequence $\{T_k:k \ge 0\}$ of stopping times such that $\{T_{k+1} - T_k:k \ge 0\}$ are independent and identically distributed; and

(ii) for every sequence of times $0 < t_1 < t_2 < ... < t_m$ ($m \ge 1$) and $k \ge 0$, the random vectors $\{X(t_1),...,X(t_m)\}$ and $\{X(T_k + t_1),...,X(T_k + t_m)\}$ have the same distribution and the processes $\{X(t):t < T_k\}$ and $\{X(T_k + t):t \ge 0\}$ are independent.

The definition of a regenerative process in discrete time is analogous. The random times $\{T_k:k \ge 0\}$ are said to be *regeneration points* for the process $\{X(t):t \ge 0\}$ and the time interval $[T_{k-1}, T_k)$ is called the kth kth *cycle* of the process.

According to Definition 1.2, every regenerative process has an embedded renewal process. The requirement that the regeneration points be stopping times means that for any fixed t the occurrence of a regeneration point prior to time t (i.e., $T_1 \le t$) depends on the evolution of the process $\{X(t):t \ge 0\}$ in the interval $(0,t]$ but not beyond time t. An irreducible and positive recurrent continuous time Markov chain (CTMC) with a finite (or countable) state space, S, is the most familiar example of a regenerative process in continuous time. The successive entrances to any fixed state $s \in S$ form a sequence of regeneration points.

Example 1.3. (Cyclic Queues With Feedback) Suppose that all service times are mutually independent and that the service times at center i are identically distributed as a positive random variable, L_i, $i = 1,2$. Let $X(t)$ be the number of jobs waiting or in service at center 2 at time t. Set $e_i =$ "service completion at center i," $i = 1,2$. Then the process $\{X(t):t \ge 0\}$ is a GSMP with finite state space, $S = \{0,1,...,N\}$ and event set, $E = \{e_1, e_2\}$.

If both L_1 and L_2 are exponentially distributed, the process $\{X(t): t \geq 0\}$ is a CTMC. The process is irreducible (in the sense that any state of the embedded jump chain is accessible from any other state) and is necessarily positive recurrent since S is finite. It follows that the successive times at which $\{X(t): t \geq 0\}$ hits a fixed state $s_0 \in S$ are regeneration points for the process and the expected time between regeneration points is finite.

If L_1 is exponentially distributed and L_2 is a positive r.v. with finite mean but otherwise arbitrary distribution, the process $\{X(t): t \geq 0\}$ is not a CTMC but is a regenerative process in continuous time. To see this, let T_n be the nth time at which there is a service completion at center 2, $n \geq 0$. The process $\{X(t): t \geq 0\}$ makes a transition to the fixed state s_0 ($0 \leq s_0 < N$) when event $e^* = e_2$ is the trigger event only if this event occurs in state $s_0 + 1$. The successive times T_n at which $X(T_n) = s_0$ are regeneration points for the process $\{X(t): t \geq 0\}$. Observe that a new service time starts at center 1 if $s_0 = N - 1$. The service time in progress at center 1 probabilistically restarts if $s_0 < N - 1$. (This is a consequence of the memoryless property of the exponential distribution. No matter when the clock for event e_1 was set, the remaining time until event e_1 triggers a state transition is exponentially distributed with the same parameter.) If $s_0 > 0$, a new service time starts at center 2. The expected time between regeneration points is finite since $\{X(T_n): n \geq 0\}$ is an irreducible, finite state discrete time Markov chain (DTMC).

Example 1.4. (System Overhead Model) Set

(1.5) $\qquad X(t) = (C_5(t), C_4(t), ..., C_1(t), S_1(t), Q_2(t))$,

where $Q_2(t)$ is the number of jobs waiting or in service at center 2, $S_1(t)$ is the job class in service at center 1, and $C_j(t)$ is the number of class j jobs in queue at center 1 at time t, $j = 1, 2, ..., 5$. ($S_1(t)$ equals 0 if there is no job in service at center 1.) Recall that the process $\{X(t): t \geq 0\}$ defined by (1.5) is a GSMP with a finite state space, S, and event set, E. The process $\{X(t): t \geq 0\}$ is a regenerative process in continuous time. To see this, let T_n be the nth time at which either (i) a class 1 service has just been completed at center 1 or (ii) after a class 1 service has been completed at center 1 with no jobs of class 2 at the center, a job of class 2 joins the queue at center 1, $n \geq 0$. The process $\{X(T_n): n \geq 0\}$ is a DTMC with a finite state space, S'. Since state $(N - 1,0,0,0,0,4,0) \in S'$ is accessible from any other state, the process $\{X(T_n): n \geq 0\}$ has a single irreducible, closed set of recurrent states. (Note that the DTMC is not irreducible; e.g., $(N - 1,0,0,0,0,5,0)$ is a transient state of the process.)

Observe that all clocks that are running at time T_n have been set (or probabilistically reset) at time T_n. The (new) clocks set at time T_n and the clock setting distributions depend on the past history of the process $\{X(t): t \geq 0\}$ only through the current state s', the trigger event, e^*, and the unique previous state s. Thus, the successive times, T_n, at which $X(T_n) = s'$, a fixed recurrent state, are regeneration points for the process $\{X(t): t \geq 0\}$. For example, $e^* = e_{11}$ is the trigger event and the previous state $s = (N - 2,1,0,0,0,1,0)$ when the process $\{X(T_n): n \geq 0\}$ hits state $s' = (N - 1,0,0,0,0,4,0)$. At these time points there is a service completion to a job of class 1 and a resumption of class 4 service at center 1, with the remaining $N - 1$ jobs waiting in queue at center 1 as jobs of class 5.

Example 1.6. (Two-server Cyclic Queues) Suppose that all service times are mutually independent. Also suppose that the service times at center i are identically distributed as a positive random variable, L_i, with finite mean but otherwise arbitrary density function that is continuous and positive on $(0, +\infty)$, $i = 1,2$. Let $Z_1(t)$ be the number of jobs waiting or in service at center 1 at time t. Also let $S_{ij}(t)$ equal one if server j at center i is busy at time t and equal zero otherwise. Set

(1.7) $X(t) = (Z_1(t), S_{11}(t), S_{12}(t), S_{21}(t), S_{22}(t))$.

Let $S^* = \{0,1,\ldots,N\} \times (\{0,1\})^4$ and event e_{ij} = "service completion by server j at center i." Then the process $\{X(t): t \geq 0\}$ is a GSMP with event set, $E = \{e_{11}, e_{12}, e_{21}, e_{22}\}$, and state space

$$S = \{(z_1, s_{11}, s_{12}, s_{21}, s_{22}) \in S^* : s_{11} + s_{12} = 2 \wedge z_1;$$
$$s_{21} + s_{22} = 2 \wedge (N - z_1)\}.$$

Without further restriction on the service time distributions, the process $\{X(t) : t \geq 0\}$ is *not* a regenerative process in continuous time. To see this, observe that for any state $s \in S$ the event set $E(s)$ contains at least two events so that there are always at least two clocks running. The assumptions on the service times imply that no two events can occur simultaneously. It follows that for any sequence $\{T_k : k \geq 0\}$ of stopping times the processes $\{X(t) : t < T_k\}$ and $\{X(T_k + t) : t \geq 0\}$ are not independent.

Example 1.8. (Queue With Scheduled Arrivals) Consider a single server queue in which the nth job arrives at time $A_n = n + L_n$, where the random variables $\{L_n : n \geq 1\}$ are independent and have a common uniform distribution on $(-1/2, 1/2)$. Also suppose that the service times $\{V_n : n \geq 1\}$ are independent and identically distributed with $1/2 < E\{V_1\} < 1$. Denote by W_n the waiting time (exclusive of service time) experienced by job n, $n \geq 0$. Then

$$W_n = [W_{n-1} + V_{n-1} - (A_n - A_{n-1})]^+$$
$$= [W_{n-1} + V_{n-1} - (1 + L_n - L_{n-1})]^+,$$

where $[x]^+$ equals x if $x \geq 0$ and equals 0 if $x < 0$. Since W_{n+1} and W_n both depend on L_n, the process $\{W_n : n \geq 0\}$ is *not* a regenerative process in discrete time.

2.2 Properties of Regenerative Processes

Let $X = \{X(t) : t \geq 0\}$ be a regenerative process in continuous time. Denote the state space of the process by S and let $\{T_k : k \geq 0\}$ be a sequence of regeneration points. Set

(2.1) $$\tau_k = T_k - T_{k-1},$$

$k \geq 1$. According to this definition τ_k is the length of the kth cycle of the regenerative stochastic process. In developing properties of regenerative processes we distinguish two cases.

2.2 Properties of Regenerative Processes

Definition 2.2. Let F be the distribution function of τ_1. The random variable τ_1 (or distribution function F) is said to be *periodic* with period $\lambda > 0$ if with probability one, τ_1 assumes values in the set $\{0,\lambda,2\lambda,...\}$ and λ is the largest such number. If there is no such λ, then τ_1 (or F) is said to be *aperiodic*.

In order for a regenerative process to have a limiting distribution when τ_1 is aperiodic, it is necessary either (i) to impose regularity conditions on the sample paths of the regenerative process, or (ii) to place restrictions of the distribution function of the time between regeneration points. Specifically, we may require that the process have right-continuous sample paths and limits from the left: for $t \geq 0$

$$X(t) = \lim_{u \downarrow t} X(u),$$

and for all $t > 0$

$$X(t-) = \lim_{u \uparrow t} X(u)$$

exists with probability one. When X is an m-dimensional stochastic process with right-continuous sample paths and limits from the left, we write $X \in D_m[0,\infty)$.

Alternatively, we may place restrictions on the distribution of the times between regeneration points. Let F_n be the n-fold convolution of the distribution function F. Denote by $\mathscr{S}$ the set of all distribution functions F such that F_n has an absolutely continuous component for some $n \geq 1$: i.e., F_n has a density function on some interval. When the distribution function F of τ_1 is an element of $\mathscr{S}$, we write $\tau_1 \in \mathscr{S}$. Most aperiodic distributions F arising in applications will be in $\mathscr{S}$.

Proposition 2.3 asserts that under mild regularity conditions, a regenerative process X has a limiting distribution ($X(t) \Rightarrow X$ as $t \to \infty$) provided that the expected time between regeneration points is finite. There is a corresponding result for the periodic case.

Proposition 2.3. Assume that τ_1 is aperiodic with $E\{\tau_1\} < \infty$. If either $X \in D_m[0,\infty)$ or $\tau_1 \in \mathscr{S}$, then $X(t) \Rightarrow X$ as $t \to \infty$.

Now suppose that τ_1 is aperiodic and for a real-valued (measurable) function f having domain, S, set

(2.4) $$r(f) = E\{f(X)\}.$$

Set

(2.5) $$Y_k(f) = \int_{T_{k-1}}^{T_k} f(X(u))du,$$

$k \geq 1$. (We always assume that the process $f(X)$ is integrable over a finite interval.) Analogously, for a regenerative process in discrete time, set

(2.6) $$Y_k(f) = \sum_{n=T_{k-1}}^{T_k-1} f(X_n).$$

Propositions 2.7 and 2.8 give fundamental properties of regenerative processes. These are the basis for the regenerative method for simulation analysis discussed in Section 2.3. Proposition 2.7 follows directly from the definition of a regenerative process.

Proposition 2.7. The sequence $\{(Y_k(f), \tau_k) : k \geq 1\}$ consists of independent and identically distributed random vectors.

Proposition 2.8 asserts that the behavior of a regenerative process within a cycle determines the limiting distribution of the process as a ratio of expected values.

Proposition 2.8. Assume that τ_1 is aperiodic with $E\{\tau_1\} < \infty$. Also assume that $E\{|f(X)|\} < \infty$. If either $X \in D_m[0, \infty)$ or $\tau_1 \in \mathscr{P}$, then

(2.9) $$E\{f(X)\} = \frac{E\{Y_1(f)\}}{E\{\tau_1\}}.$$

There is an analogous *ratio formula* when τ_1 is periodic.

2.3 The Regenerative Method for Simulation Analysis

We have seen that in the presence of certain regularity conditions, a regenerative stochastic process $\{X(t):t \geq 0\}$ has a limiting distribution provided that the expected time between regeneration points is finite. Furthermore, the regenerative structure ensures that the behavior of the regenerative process in a cycle determines the limiting distribution of the process as a ratio of expected values. A consequence of these results is that a strongly consistent point estimate and asymptotic confidence interval for the expected value of a general (measurable) function of the limiting random variable can be obtained by observing of a finite portion of a single sample path of the process. This comprises the regenerative method and is accomplished by simulating the process in cycles and measuring quantities determined by the individual cycles.

Where applicable, this regenerative method is attractive because it provides point and interval estimates having desirable properties. There are, however, other considerations. The classical alternative entails selecting an initial state for the process, running the simulation for an initial period of time (and discarding this "initial transient"), and then observing the process ("in steady state") for an additional period of time from which point estimates are obtained. In general, no confidence interval is available, nor is there any guidance on the selection of the initial state. Moreover, the determination of a suitable initial period of time and a suitable additional period of time is often nontrivial and likely to require sophisticated statistical techniques. With the regenerative method, these difficulties to a large extent are avoidable.

Recall from Definition 1.2 that a real (possibly vector-valued) stochastic process $\{X(t):t \geq 0\}$ having state space, S, is a regenerative process in continuous time provided that: (i) there exists a sequence $\{T_k:k \geq 0\}$ of stopping times that form a renewal process; and (ii) for every sequence of times $0 < t_1 < t_2 < ... < t_m$ ($m \geq 1$) and $k \geq 0$, the random vectors $\{X(t_1),...,X(t_m)\}$ and $\{X(T_k + t_1),...,X(T_k + t_m)\}$ have the same distribution and the processes $\{X(t): t < T_k\}$ and $\{X(T_k + t): t \geq 0\}$ are independent. We assume that

$$\tau_1 = T_1 - T_0$$

is aperiodic and that for a real-valued (measurable) function f having domain S, the goal of the simulation is the estimation of the quantity

$$r(f) = E\{f(X)\}.$$

A strongly consistent point estimate and aymptotic confidence interval for $r(f)$ can be obtained by observing a finite portion of a single sample path of the regenerative process $\{X(t): t \geq 0\}$. We assume throughout that the regenerative process and the function f are such that the ratio formula for $r(f)$ holds:

$$r(f) = \frac{E\{Y_k(f)\}}{E\{\tau_k\}},$$

where $Y_k(f)$ and τ_k are given by (2.1) and (2.5), respectively. Set

(3.1) $$Z_k(f) = Y_k(f) - r(f)\tau_k$$

and observe that the kth cycle of the regenerative process $\{X(t): t \geq 0\}$ completely determines the quantity $Z_k(f)$. The sequence $\{Z_k(f): k \geq 1\}$ defined by (3.1) consists of i.i.d. random variables and the ratio formula for $r(f)$ implies that $E\{Z_k(f)\} = 0$. Set

(3.2) $$\sigma^2 = \text{var}(Z_1(f)) = \text{var}(Y_1(f) - r(f)\tau_1).$$

Writing

$$E\{(Z_1(f))^2\} = E\left\{[(Y_1(f) - E\{Y_1(f)\}) - r(f)(\tau_1 - E\{\tau_1\})]^2\right\},$$

it follows that

(3.3) $$\sigma^2 = \text{var}(Y_1(f)) - 2r(f)\,\text{cov}(Y_1(f), \tau_1) + (r(f))^2\,\text{var}(\tau_1).$$

(We always assume that $0 < \sigma^2 < \infty$. The case $\sigma^2 = 0$ is degenerate and $\sigma^2 < \infty$ for most finite state processes. In some queueing systems, however, additional finite higher moment conditions on service and interarrival times are needed to ensure that $\sigma^2 < \infty$.)

Now fix n, the number of cycles, and let $\bar{Y}(n)$, $\bar{\tau}(n)$, $s_{11}(n)$, $s_{22}(n)$, and $s_{12}(n)$ be the usual unbiased point estimates of $E\{Y_1(f)\}$, $E\{\tau_1\}$, var $(Y_1(f))$, var (τ_1), and cov $(Y_1(f), \tau_1)$, respectively:

$$\bar{Y}(n) = \frac{1}{n} \sum_{k=1}^{n} Y_k(f),$$

$$\bar{\tau}(n) = \frac{1}{n} \sum_{m=1}^{n} \tau_m,$$

$$s_{11}(n) = \frac{1}{n-1} \sum_{k=1}^{n} (Y_k(f) - \bar{Y}(n))^2,$$

$$s_{22}(n) = \frac{1}{n-1} \sum_{k=1}^{n} (\tau_k - \bar{\tau}(n))^2,$$

and

$$s_{12}(n) = \frac{1}{n-1} \sum_{k=1}^{n} (Y_k(f) - \bar{Y}(n))(\tau_k - \bar{\tau}(n)).$$

As a consequence of the strong law of large numbers for i.i.d. sequences of random variables, the point estimate

(3.4) $$\hat{r}(n) = \frac{\bar{Y}(n)}{\bar{\tau}(n)}$$

converges with probability one to $r(f)$ and

(3.5) $$s(n) = \left\{ s_{11}(n) - 2\hat{r}(n) \, s_{12}(n) + (\hat{r}(n))^2 \, s_{22}(n) \right\}^{1/2}$$

converges with probability one to σ as $n \to \infty$. Thus, by definition, $\hat{r}(n)$ and $s(n)$ are *strongly consistent* point estimates.

The construction of asymptotic confidence intervals for $r(f)$ rests on a particular central limit theorem (c.l.t.).

Proposition 3.6. Assume that $0 < \sigma^2 < \infty$. Then

$$(3.7) \qquad \frac{n^{1/2}\{\hat{r}(n) - r(f)\}}{\sigma/E\{\tau_1\}} \Rightarrow N(0,1)$$

as $n \to \infty$.

Proof. The standard c.l.t. for i.i.d. mean 0, finite variance random variables implies that

$$(3.8) \qquad \frac{1}{\sigma n^{1/2}} \sum_{k=1}^{n} Z_k(f) \Rightarrow N(0,1)$$

as $n \to \infty$. This can be rewritten as

$$(3.9) \qquad \frac{n^{1/2}\{\hat{r}(n) - r(f)\}}{(\sigma/E\{\tau_1\})(E\{\tau_1\}/\bar{\tau}(n))} \Rightarrow N(0,1)$$

as $t \to \infty$. The strong law of large numbers guarantees that

$$\frac{E\{\tau_1\}}{\bar{\tau}(n)} \Rightarrow 1$$

as $n \to \infty$. Lemma 1.6 of Appendix 1 applies to this situation, and hence

$$\left(X_n, \frac{E\{\tau_1\}}{\bar{\tau}(n)}\right) \Rightarrow (N(0,1), 1)$$

as $n \to \infty$, where X_n denotes the left hand side of (3.9). Now apply the continuous mapping theorem using the mapping h given by $h(x,y) = xy$ to conclude that

$$(3.10) \qquad X_n \cdot \frac{E\{\tau_1\}}{\bar{\tau}(n)} \Rightarrow N(0,1) \cdot 1$$

as $n \to \infty$. Since $N(0,1) \cdot 1$ has the same distribution as $N(0,1)$, (3.10) is the same as (3.7). □

2.3 The Regenerative Method for Simulation Analysis

An asymptotic confidence interval for $r(f)$ can be obtained from (3.7) but in general, of course, the "standard deviation constant" $\sigma/E\{\tau_1\}$ is not known. The most straightforward estimate for $\sigma/E\{\tau_1\}$ is $s(n)/\bar{\tau}(n)$, and the strong law of large numbers ensures that $\sigma/s(n) \Rightarrow 1$ as $n \to \infty$. The same argument that leads to (3.7) yields the c.l.t.

$$\text{(3.11)} \qquad \frac{n^{1/2}\{\hat{r}(n) - r(f)\}}{s(n)/\bar{\tau}(n)} \Rightarrow N(0,1)$$

as $n \to \infty$.

Let Φ be the the distribution function of $N(0,1)$ and set $z_{1-\gamma} = \Phi^{-1}(1-\gamma)$, $0 < \gamma < 1/2$. It follows from (3.11) that

$$\text{(3.12)} \qquad \left[\hat{r}(n) - \frac{z_{1-\gamma}\, s(n)}{\bar{\tau}(n)\, n^{1/2}},\ \hat{r}(n) + \frac{z_{1-\gamma}\, s(n)}{\bar{\tau}(n)\, n^{1/2}} \right]$$

is an *asymptotic* $100(1-2\gamma)\%$ *confidence interval* for $r(f)$:

$$\lim_{n \to \infty} P\{r(f) \in \hat{I}(n)\} = 1 - 2\gamma,$$

where $\hat{I}(n)$ is the interval in (3.12). Thus, when n is large the interval $\hat{I}(n)$ contains the unknown constant $r(f)$ approximately $100(1-2\gamma)\%$ of the time. For 90% ($\gamma = .05$) confidence intervals, $z_{1-\gamma} = 1.645$; for 95% ($\gamma = .025$) confidence intervals, $z_{1-\gamma} = 1.96$. The confidence interval has random endpoints and is symmetric about the point estimate $\hat{r}(n)$. The half length of the interval is $n^{-1/2}$ times a multiple ($z_{1-\gamma}$) of the estimate of the standard deviation constant $\sigma/E\{\tau_1\}$. Thus as n increases, the length of the interval converges to 0 and the midpoint converges to the true value.

Algorithm 3.13. (Regenerative Method)
1. Select a sequence, $\{T_k : k \geq 0\}$, of regeneration points.
2. Simulate the regenerative process and observe a fixed number, n, of cycles defined by the $\{T_k : k \geq 0\}$.
3. Compute $\tau_m = T_m - T_{m-1}$, the length of the mth cycle,

and the quantity

$$Y_m(f) = \int_{T_{m-1}}^{T_n} f(X(u))du.$$

4. Form the point estimate

$$\hat{r}(n) = \frac{\bar{Y}(n)}{\bar{\tau}(n)}.$$

5. Form the asymptotic $100(1 - 2\gamma)\%$ confidence interval

$$\left[\hat{r}(n) - \frac{z_{1-\gamma}}{\bar{\tau}(n)} \frac{s(n)}{n^{1/2}}, \hat{r}(n) + \frac{z_{1-\gamma}}{\bar{\tau}(n)} \frac{s(n)}{n^{1/2}} \right].$$

There is an analogous procedure for obtaining asymptotic confidence intervals from a fixed length simulation of a regenerative process. The procedure for a regenerative simulation of the process $\{X(u): 0 \leq u \leq t\}$ the same as that of Algorithm 3.13 except that statistics are computed only for the random number, $n(t)$, of cycles completed by time t. Asymptotic confidence intervals for $r(f)$ are based on a c.l.t. that corresponds to (3.7):

(3.14)
$$\frac{t^{1/2} \{\hat{r}(n(t)) - r(f)\}}{\sigma/(E\{\tau_1\})^{1/2}} \Rightarrow N(0,1)$$

as $t \to \infty$.

Example 3.15. (System Overhead Model) The process $\{X(t): t \geq 0\}$ defined by (1.5) is a regenerative process in continuous time and $X(t) \Rightarrow X$ as $t \to \infty$. Denote the state space of the process by S. Let f_j be a real-valued function such that $f_j(c_5,...,c_1,s_1,q_2) = 1_{\{s_1=j\}}$ for $(c_5,...,c_1,s_1,q_2) \in S$. and set $r(f_j) = E\{f_j(X)\}$. (The quantity $r(f_j)$ is the limiting probability that center 1 provides service to jobs of class j, $j = 1,2,...,5$.)

Let $\{T_k: k \geq 0\}$ be a sequence of regeneration points for the process $\{X(t): t \geq 0\}$. Then $\tau_k = T_k - T_{k-1}$ is the length of the kth cycle and $Y_k(f_j)$ is the total amount of time in the kth cycle that center 1 provides service to jobs of class j. Propositions 2.7 and 2.8 ensure that the pairs of random variables $\{(Y_k(f_j),\tau_k):k \geq 1\}$ are i.i.d. and

$$r(f_j) = \frac{E\{Y_1(f_j)\}}{E\{\tau_1\}}.$$

Based on n cycles, a strongly consistent point estimate for $r(f_j)$ is

$$\hat{r}_j(n) = \frac{\bar{Y}_j(n)}{\bar{\tau}(n)}$$

and an asymptotic $100(1 - 2\gamma)\%$ confidence interval is

$$\left[\hat{r}_j(n) - \frac{z_{1-\gamma}s_j(n)}{\bar{\tau}(n)\, n^{1/2}},\; \hat{r}_j(n) + \frac{z_{1-\gamma}s_j(n)}{\bar{\tau}(n)\, n^{1/2}}\right].$$

The quantity $s_j^2(n)$, where

$$s_j(n) = \left\{s_{11}(n) - 2\hat{r}_j(n)\, s_{12}(n) + (\hat{r}_j(n))^2 s_{22}(n)\right\}^{1/2},$$

is a strongly consistent point estimate for $\sigma_i^2 = \mathrm{var}\,(Y_1(f_j) - r(f_j)\tau_1)$ and $\bar{Y}_j(n)$, $\bar{\tau}(n)$, $s_{11}(n)$, $s_{22}(n)$, and $s_{12}(n)$ are the usual unbiased point estimates of $E\{Y_1(f_j)\}$, $E\{\tau_1\}$, $\mathrm{var}\,(Y_1(f_j))$, $\mathrm{var}\,(\tau_1)$, and $\mathrm{cov}\,(Y_1(f_j),\tau_1)$, respectively. Asymptotic confidence intervals for $r(f_j)$ are based on the c.l.t.

$$\frac{n^{1/2}\{\hat{r}_j(n) - r(f_j)\}}{s_j(n)/\bar{\tau}(n)} \Rightarrow N(0,1)$$

as $n \to \infty$.

2.4 Implementation Considerations

In applications of the regenerative method some care must be taken in the computation of the quantities $s_{11}(n)$, $s_{22}(n)$, and $s_{12}(n)$ in (3.5). For example, the computation of $s_{11}(n)$ according to

$$s_{11}(n) = \frac{1}{n-1} \sum_{k=1}^{n} (Y_k(f))^2 - n(\bar{Y}(n))^2$$

requires only one pass through the data $Y_1(f), Y_2(f), \ldots, Y_n(f)$. This form is numerically unstable and error due to round-off and cancellation can be substantial. There is a numerically stable one-pass updating algorithm that can be used to compute $s_{11}(n)$. Set $s_{11}^{(1)} = 0$ and compute

$$s_{11}^{(j+1)} = s_{11}^{(j)} + \frac{\left[\sum_{k=1}^{j} Y_k(f) - jY_{j+1}(f)\right]^2}{j(j+1)},$$

$j = 1, 2, \ldots, n - 1$. Then set

$$s_{11}(n) = s_{11}^{(n)}/(n-1).$$

The sample variance $s_{22}(n)$ can be computed from the observations $\tau_1, \tau_2, \ldots, \tau_n$ in the same manner.

There is an analogous one-pass method that can be used to compute the sample covariance, $s_{12}(n)$. Set $s_{12}^{(1)} = 0$ and compute

$$s_{12}^{(j+1)} = s_{12}^{(j)} + \frac{\left[\sum_{k=1}^{j} Y_k(f) - jY_{j+1}(f)\right]\left[\sum_{k=1}^{j} \tau_k - j\tau_{k+1}\right]}{j(j+1)},$$

$j = 1, 2, \ldots, n - 1$. Then set

$$s_{12}(n) = s_{12}^{(n)}/(n-1).$$

Given $s_{11}(n)$, $s_{22}(n)$, and $s_{12}(n)$, the quantity $s(n)$ can be computed from (3.5). The main deficiency of this method arises from possible instability in the calculation of $s_{12}(n)$ with a resulting round-off and cancellation error. An alternative stable two-pass method for computing $s^2(n)$ is

$$s^2(n) = \frac{1}{n-1} \sum_{k=1}^{n} \left(Y_k(f) - \hat{r}(n)\tau_k\right)^2,$$

where $\hat{r}(n)$ is computed on the first pass.

A regenerative process can possess more than one sequence of regeneration points and cycles associated with one such sequence can be very much longer than cycles associated with another.

Proposition 4.4 asserts that with high probability the resulting confidence intervals are of the same length, provided that the length of the simulation run is large. Let $\{T_k(i):k \geq 0\}$ be two sequences of regeneration points for the regenerative process $\{X(t):t \geq 0\}$, $i = 1,2$. (As before, we assume that the regenerative process and the function f are such that ratio formulas for $r(f)$ hold.) Set $\tau_k(i) = T_k(i) - T_{k-1}(i)$,

$$Y_{k,i}(f) = \int_{T_{k-1}(i)}^{T_k(i)} f(X(u))du,$$

and $Z_{k,i}(f) = Y_{k,i}(f) - r(f)\tau_k(i)$. Each sequence $\{Z_{k,i}(f):k \geq 1\}$ consists of i.i.d. random variables and $E\{Z_{k,i}(f)\} = 0$. Set $\sigma_i^2 = \text{var}(Z_{k,i}(f))$.

Lemma 4.1. Suppose that $E\{\tau_1(i)\} < \infty$, $i = 1,2$. Also suppose that $E\{Y_{1,1}(|f|)\} < \infty$ and $E\{(Y_{1,1}(|f - r(f)|))^2\} < \infty$. Then

(4.2) $$\frac{\sigma_1^2}{E\{\tau_1(1)\}} = \frac{\sigma_2^2}{E\{\tau_1(2)\}}.$$

This lemma follows from the convergence of types theorem and two c.l.t.'s:

(4.3) $$\frac{t^{1/2}\left\{\int_0^t f(X(u))du - r(f)t\right\}}{\sigma_i/(E\{\tau_1(i)\})^{1/2}} \Rightarrow N(0,1)$$

as $t \to \infty$. Now consider a simulation of fixed length t and denote by $n_i(t)$ the number of cycles (defined by $\{T_k(i):k \geq 0\}$) completed in $[0,t)$. For fixed γ let $I_i(t)$ be the length of the asymptotic $100(1 - 2\gamma)\%$ confidence intervals for $r(f)$ obtained from (4.3).

Proposition 4.4. Under the conditions of Lemma 4.1,

$$\lim_{t \to \infty} \frac{I_1(t)}{I_2(t)} = 1$$

with probability one.

Proof. By the strong laws for renewal processes and for partial sums,

$$\frac{n_i(t)}{t} \to \frac{1}{E\{\tau_1(i)\}}$$

and $s_i^2(n_i(t)) \to \sigma_i^2$ with probability one as $t \to \infty$. The definitions of $n_i(t)$ and $I_i(t)$ together with (4.1) yield

$$(4.5) \qquad (n_i(t))^{1/2} I_i(t) = \frac{2 z_{1-\gamma} s_i(n_i(t))}{(\bar{\tau}_1(n_i(t)))^{1/2}}.$$

Now let $t \to \infty$ and observe that

$$(4.6) \qquad (n_i(t))^{1/2} I_i(t) \Rightarrow \frac{2 z_{1-\gamma} \sigma_i}{(E\{\tau_1(i)\})^{1/2}}$$

as $t \to \infty$. The proposition now follows from Lemma 4.1. □

The c.l.t. of (3.7) implies that the half length of the confidence interval based on n cycles is $z_{1-\gamma} \sigma/(E\{\tau_1\} n^{1/2})$. It follows that the number, $n(\gamma, \delta)$, of cycles required to obtain an asymptotic $100(1 - 2\gamma)\%$ confidence interval for $r(f)$ whose half length is $100\delta\%$ of $r(f)$ is (approximately)

$$(4.7) \qquad n(\gamma, \delta) = \left(\frac{z_{1-\gamma}}{\delta}\right)^2 \left(\frac{\sigma}{r(f) E\{\tau_1\}}\right)^2.$$

The first factor, $(z_{1-\gamma}/\delta)^2$, is independent of the system being simulated. From the second factor, $[\sigma/(r(f) E\{\tau_1\})]^2$, it is apparent that some systems are inherently more difficult to simulate than others. This quantity provides a good measure of the amount of simulation required for a fixed level of precision. An estimate of $[\sigma/(r(f) E\{\tau_1\})]^2$ obtained from a pilot run can be used to determine the length of the final simulation run. Given this estimate, a value of n can be computed form (4.7) and used as the sample size n for the simulation. This method has been called a *two-stage procedure*. It is clear from (4.7) that every simulation involves a trade-off between degree of confidence, length of confidence interval, and run length (sample size, n).

It is also possible to use a *sequential procedure* in which the number of cycles is a random variable determined dynamically in the course of the simulation. The value of the random variable n is the least integer m such that

$$m \geq \left(\frac{z_{1-\gamma}}{\delta}\right)^2 \left(\frac{s(m)}{\hat{r}(m)\bar{\tau}(m)}\right)^2.$$

2.5 Theoretical Values for Discrete Time Markov Chains

When developing simulation methodology it is important to assess the statistical efficiency of proposed estimation procedures. In this section we develop computational methods for discrete time Markov chains. These methods make it possible to obtain theoretical values for variance constants entering into c.l.t.'s used to form confidence intervals in regenerative simulations.

Let $\{X_k : k \geq 0\}$ be an irreducible DTMC with finite state space, $S = \{0, 1, ..., N\}$, and one-step transition matrix, $P = \{p_{ij} : i, j \in S\}$. For this chain denote the n-step transition probability from state i to state j by p_{ij}^n and recall that

$$P^n = \{p_{ij}^n : i, j \in S\}.$$

Throughout this section we use the following notation. For a fixed state $i \in S$, $P_i\{\cdot\}$ denotes the conditional probability associated with starting the chain in state i, and $E_i\{\cdot\}$ denotes the corresponding conditional expectation. For $j \in S$ and $n \geq 1$, we denote by $T_n(j)$ the nth entrance time of $\{X_k : k \geq 0\}$ to state j; e.g.,

$$T_1(j) = \min\{k \geq 1 : X_k = j\}.$$

Set $\tau_1(j) = T_1(j)$ and $\tau_n(j) = T_n(j) - T_{n-1}(j)$, $n > 1$. This notation is consistent with that introduced in Section 2.2 for regenerative processes. Note that $\{T_n(j) : n \geq 1\}$ is a (possibly delayed) renewal process since a finite state, irreducible CTMC is necessarily positive recurrent and therefore returns to every state $j \in S$ infinitely often with probability one. If $X_0 = j$, the process $\{T_n(j):n \geq 1\}$ is an ordinary renewal process.

We consider vectors such as $(v(0),v(1),...,v(N))$ to be column vectors. Real-valued functions such as f and g, having domain S, are viewed in this way and denoted by $\mathbf{f}$ and $\mathbf{g}$. In this context the symbol $E\{\cdot\}$ denotes the vector

$$(E_0\{\cdot\},E_1\{\cdot\},...,E_N\{\cdot\}).$$

In addition (for vectors $\mathbf{u}$ and $\mathbf{v}$) the symbol $\mathbf{u} \circ \mathbf{v}$ denotes the Hadamard product of vectors

$$(u(0)v(0),u(1)v(1),...,u(N)v(N)).$$

The component of the vector $\mathbf{v}$ corresponding to state j is denoted by $[\mathbf{v}]_j$. For a matrix $A = (a_0,a_1,...,a_m)$, set

$$\mathbf{u} \circ A = A \circ \mathbf{u} = (\mathbf{u} \circ a_0, \mathbf{u} \circ a_1,...,\mathbf{u} \circ a_m).$$

Finally, for a matrix $B = (b_0,b_1,...,b_m)$, set

$$A \circ B = (a_0 \circ b_0, a_1 \circ b_1,...,a_m \circ b_m).$$

We now show how to compute the quantities $E_i\{Y_1(f)\}$ and $E_i\{Y_1(f)Y_1(g)\}$ for general real-valued functions f and g having domain S and $i \in S$. We restrict attention to cycles regenerative process $\{X_k:k \geq 0\}$ formed by the successive entrances to state 0, and henceforth suppress the 0 in the notation $T_n(0), \tau_n(0)$, etc. Note that this is no real restriction, and that equally well we could choose any other state $i \in S$. For $i,j \in S$ and $n = 0,1,...$, set

$$_0p_{ij}^n = P_i\{\tau_1 > n, X_n = j\}$$

and

$$_0P^n = \{_0p_{ij}^n : i,j \in S\}.$$

We obtain $_0P^1 = {_0P}$ from P by setting the 0-column of P equal to 0. It is easy to see that $_0P^n$ is the product of n copies of $_0P$, and that $_0p_{i0}^n = 0$ for all $n \geq 1$.

2.5 Theoretical Values for Discrete Time Markov Chains

For any real-valued function f with domain S, set

$$Y_1(f) = \sum_{k=0}^{T_1-1} f(X_k)$$

and

$$Z_1(f) = Y_1(f) - r(f)\tau_1.$$

Proposition 5.1. For an irreducible, finite state discrete time Markov chain with transition matrix P,

(5.2) $$E\{Y_1(f)\} = (I - {}_0P)^{-1}f$$

and

(5.3) $$E\{Y_1(f)Y_1(g)\} = (I - {}_0P)^{-1}h,$$

where

$$h = f \circ E\{Y_1(g)\} + g \circ E\{Y_1(f)\} - f \circ g.$$

Proof. First write

(5.4) $$Y_1(f) = \sum_{n=0}^{\infty} f(X_n) 1_{\{T_1>n\}} = \sum_{n=0}^{\infty} 1_{\{T_1>n\}} \sum_{j \in S} f(j) 1_{\{X_n=j\}}.$$

Since

$$E_i\{1_{\{T_1>n\}} 1_{\{X_n=j\}}\} = {}_0p_{ij}^n,$$

we have

$$E_i\{Y_1(f)\} = \sum_{n=0}^{\infty} \sum_{j \in S} {}_0p_{ij}^n f(j) = \sum_{j \in S} (I - {}_0P)_{ij}^{-1} f(j)$$

which is equivalent to (5.2). The various interchanges of Σ and E_i are justified since

$$\sum_{n=0}^{\infty} \sum_{j \in S} {}_0p_{ij}^n |f(j)| < \infty.$$

To establish (5.3), we write $Y_1(g)$ in the form given by (5.4). The product $Y_1(f)Y_1(g)$ can be written as

(5.5) $$Y_1(f)Y_1(g) = \left(\sum_{n=0}^{\infty} a_n\right)\left(\sum_{m=0}^{\infty} b_m\right)$$

$$= \left(\sum_{n=0}^{\infty} a_n\right)\left(\sum_{m=n}^{\infty} b_m\right) + \left(\sum_{n=0}^{\infty} a_n\right)$$

where the sequences $\{a_n : n \geq 0\}$ and $\{b_m : m \geq 0\}$ are identified by (5.4) in the obvious manner. Next we observe that

$$E_i\{1_{\{T_1 > n\}} 1_{\{X_n = j\}} 1_{\{T_1 > m+n\}} 1_{\{X_{n+m} = k\}}\} = {}_0p_{ij}^n \, {}_0p_{jk}^m.$$

Taking the first term on the right hand side of (5.5) we find that

$$E_i\left\{\left(\sum_{n=0}^{\infty} a_n\right)\left(\sum_{m=0}^{\infty} b_{n+m}\right)\right\}$$

$$= \sum_{n=0}^{\infty} \sum_{m=0}^{\infty} \sum_{j \in S} \sum_{k \in S} f(j) \, g(k) \, {}_0p_{ij}^n \, {}_0p_{jk}^m$$

$$= \sum_{n=0}^{\infty} \sum_{j \in S} f(j) \, {}_0p_{ij}^n \sum_{m=0}^{\infty} \sum_{k \in S} g(k) \, {}_0p_{jk}^m$$

$$= \sum_{n=0}^{\infty} \sum_{j \in S} (f \circ E\{Y_1(g)\})(j) \, {}_0p_{ij}^n$$

$$= \sum_{j \in S} (I - {}_0P)_{ij}^{-1} (f \circ E\{Y_1(g)\})(j).$$

The second term on the right hand side of (5.5) can be handled in exactly the same way by simply interchanging f and g since

$$\sum_{n=0}^{\infty} \sum_{m=0}^{\infty} a_n b_m = \sum_{m=0}^{\infty} \sum_{n=m}^{\infty} b_m a_n.$$

Finally, the last term on the right hand side of (5.5) can be obtained from (5.2) with f replaced by $f \circ g$. These three terms combined yield (5.3). □

Corollary 5.6. For an irreducible, finite state discrete time Markov chain with transition matrix P,

(5.7) $$E\{\tau_1\} = (I - {_0P})^{-1} e$$

and

(5.8) $$E\{Z_1^2(f)\} = (I - {_0P})^{-1} h,$$

where $e = (1,\ldots,1)$ and

$$h = 2(f - r(f)e) \circ E\{Y_1(f - r(f)e)\} - (f - r(f)e) \circ (f - r(f)e).$$

Proof. Using the fact that $\tau_1 = Y_1(e)$, (5.7) follows from Proposition 5.1. Alternatively,

$$E_i\{T_1\} = \sum_{n=0}^{\infty} P_i\{T_1 > n\} = \sum_{n=0}^{\infty} \sum_{j=1}^{N} {_0p_{ij}^n} = [(I - {_0P})^{-1} e]_i.$$

To establish (5.8), observe that

$$Z_1(f) = \sum_{n=0}^{T_1-1} \{f(X_n) - r(f)\} = Y_1(f - r(f)e)$$

and apply (5.3). □

2.6 Theoretical Values for Continuous Time Markov Chains

Now let $\{X(t): t \geq 0\}$ be a CTMC with finite state space, $S = \{0, 1, \ldots, N\}$, and transition matrix, $P(t) = \{p_{ij}(t) : i, j \in S\}$. Recall that in a CTMC, the infinitesimal generator, $Q = \{q_{ij} : i, j \in S\}$, is the given data and $Q = P'(0)$. The exponentially distributed holding time in any state $i \in S$ has mean $q_i^{-1} = -q_{ii}^{-1}$. For all $i \in S$, we assume that $0 < q_i < \infty$ so that all states are stable and nonabsorbing. In addition we assume that

$$\sum_{j=0}^{N} q_{ij} = 0$$

so that, starting from any state $i \in S$, the CTMC makes a transition to a next state $j \in S$. The elements of the jump matrix $\boldsymbol{R} = \{r_{ij}: i,j \in S\}$ of the CTMC are

$$r_{ij} = \begin{cases} q_{ij}/q_i & \text{if } j \neq i \\ 0 & \text{if } j = i \end{cases}$$

We assume that $\boldsymbol{R}$ is irreducible; this is equivalent to the CTMC being irreducible (and therefore positive recurrent). As before, we let $P_i\{\cdot\}$ and $E_i\{\cdot\}$ denote the conditional probability and conditional expectation associated with starting in state $i \in S$. For $j \in S$ and $n \geq 1$, let $T_n(j)$ be the nth entrance time of $\{X(t): t \geq 0\}$ to state j; e.g.,

$$T_1(j) = \inf\{s > 0: X(s-) \neq j, X(s) = j\}.$$

Then set $\tau_1(j) = T_1(j)$ and

$$\tau_n(j) = T_n(j) - T_{n-1}(j),$$

$n > 1$.

We now consider the computation of $E_i\{Y_1(f)\}$ and $E_i\{Y_1(f)Y_1(g)\}$ for real valued functions f and g with domain S, $i \in S$. As in the case of DTMC's, we restrict attention to regenerative cycles formed by the successive entrances to state 0, and suppress the 0 in our notation. Set

$$_0P_{ij}(t) = P_i\{\tau_1 > t, X(t) = j\}$$

and

$$_0P(t) = \{_0P_{ij}(t): i,j \in S\}.$$

Next construct the matrix $_0\boldsymbol{R}^n$ from $\boldsymbol{R}$ in the same manner as $_0\boldsymbol{P}^n$ was constructed from $\boldsymbol{P}$ in the discrete time case, $n \geq 0$.

For a real-valued function f having domain S, set

$$Y_1(f) = \int_0^{T_1} f(X(s))\,ds$$

and

$$Z_1(f) = Y_1(f) - r(f)\tau_1.$$

Proposition 6.1. For an irreducible, finite state continuous time Markov chain with jump matrix R and vector q of rate parameters for holding times,

(6.2) $\quad E\{Y_1(f)\} = E\left\{\int_0^\infty {}_0P(t)f\,dt\right\} = (I - {}_0R)^{-1}(f \circ q^{-1})$

and

(6.3) $\quad E\{Y_1(f)Y_1(g)\} = E\left\{\int_0^\infty {}_0P(t)h\,dt\right\} = (I - {}_0R)^{-1}(h \circ q^{-1}),$

where

$$h = f \circ E\{Y_1(g)\} + g \circ E\{Y_1(f)\}$$

and the column vector $q^{-1} = (q_0^{-1}, q_1^{-1}, \ldots, q_N^{-1})$.

Proof. As in the discrete time case first write

$$Y_1(f) = \int_0^\infty f(X(t))\, 1_{\{T_1 > t\}}\,dt = \int_0^\infty 1_{\{T_1 > t\}} \sum_{j \in S} f(j)\, 1_{\{X(t)=j\}}\,dt.$$

Taking the expectation with respect to E_i on both sides and using

$$E_i\{1_{\{T_1 > t\}} 1_{\{X(t)=j\}}\} = {}_0P_{ij}(t)$$

yields

$$E_i\{Y_1(f)\} = \int_0^\infty \sum_{j \in S} f(j)\, {}_0P_{ij}(t)\,dt,$$

which is the first expression in (6.2). Again the interchanges of

E_i, $\int$, and Σ are justified since $E_i\{T_j\} < \infty$ and

$$\int_0^\infty \sum_{j \in S} |f(j)| \, _0p_{ij}(t) dt < \infty.$$

To obtain the second expression in (6.2), set $\zeta_0 = 0$ and let ζ_n be the nth time at which the process makes a state transition, $n \geq 1$. Since $\{X(t): t \geq 0\}$ is irreducible and S is finite, $\zeta_n \to \infty$ with probability one for each initial state $i \in S$. As

$$[0, +\infty) = \bigcup_{n=0}^\infty [\zeta_n, \zeta_{n+1}),$$

we can write

(6.4) $$E_i\left\{\int_0^\infty f(X(t)) \, 1_{\{T_1 > t\}} \, dt\right\}$$

$$= \sum_{n=0}^\infty E_i\left\{\int_{\zeta_n}^{\zeta_{n+1}} f(X(t)) \, 1_{\{T_1 > t\}} \, dt\right\}.$$

Let $\{X_n : n \geq 0\}$ be the embedded jump chain of the CTMC and denote by $\delta_n(j)$ the length of the nth j-cycle. Then

(6.5) $$E_i\left\{\int_{\zeta_n}^{\zeta_{n+1}} f(X(t)) \, 1_{\{T_1 > t\}} dt\right\}$$

$$= E_i\{f(X_n) \, 1_{\{\delta_1 > n\}} \, (\zeta_{n+1} - \zeta_n)\}$$

$$= E_i\{f(X_n) \, 1_{\{\delta_1 > n\}} \, E\{\zeta_{n+1} - \zeta_n | X_n\}\}$$

$$= \sum_{j \in S} {_0r(j)}_{ij}^n \, (f(j) q_j^{-1})$$

$$= [_0R^n (f \circ q^{-1})]_i.$$

2.6 Theoretical Values for Continuous Time Markov Chains

Combining (6.4) and (6.5) we have

$$E\{Y_1(f)\} = \sum_{n=0}^{\infty} {}_0R^n (f \circ q^{-1}) = (I - {}_0R)^{-1} (f \circ q^{-1}),$$

the second part of (6.2).

To establish (6.3) we begin by writing

(6.6) $\quad Y_1(f)Y_1(g) = \left(\int_{t=0}^{\infty}\right) \left(\int_{s=0}^{\infty}\right)$

$$= \left(\int_{t=0}^{\infty}\right) \left(\int_{s=t}^{\infty}\right) + \left(\int_{t=0}^{\infty}\right) \left(\int_{s=0}^{t}\right).$$

Then we take the first term on the right hand side in (6.6) and write it as

$$\left(\int_{t=0}^{\infty}\right) \left(\int_{s=t}^{\infty}\right)$$

$$= \int_{t=0}^{\infty} \int_{s=0}^{\infty} f(X(t)) \, 1_{\{T_1 > t\}} \, g(X(t+s)) \, 1_{\{T_1 > t+s\}} ds \, dt$$

$$= \int_{t=0}^{\infty} 1_{\{T_1 > t\}} \sum_{j \in S} f(j) \, 1_{\{X(t) = j\}} dt$$

$$\int_{s=0}^{\infty} 1_{\{T_1 > t+s\}} \sum_{k \in S} g(k) \, 1_{\{X(t+s) = k\}} ds.$$

Now take expectations on both sides and obtain

$$E_i \left\{ \int_{t=0}^{\infty} \int_{s=t}^{\infty} \right\} = \int_0^{\infty} \sum_{j \in S} f(j) {}_0p_{ij}(t) dt \int_0^{\infty} \sum_{k \in S} g(k) {}_0p_{jk}(s) ds$$

$$= \int_0^{\infty} \sum_{j \in S} {}_0p_{ij}(t) \, [f(j) E_j\{Y_1(g)\}] ds.$$

This term is exactly the contribution to (6.6) from the first term of h. The second term on the right hand side of the equation can be handled in exactly the same way. The second expression in (6.6) follows from (6.4). □

Corollary 6.7. For an irreducible, finite state continuous time Markov chain with jump matrix R and vector q of rate parameters for holding times,

$$(6.8) \qquad E\{T_1\} = (I - {_0P})^{-1} q^{-1}$$

and

$$(6.9) \qquad E\{Z_1^2(f)\} = (I - {_0P})^{-1} (h \circ q^{-1}),$$

where

$$h = 2(f - r(f)e) \circ E\{Y_1(f - r(f)e)\}.$$

Proof. The expression in (6.8) follows from (6.2) and the fact that $T_1 = Y_1(e)$. Similarly, (6.9) follows from

$$Z_1(f) = \int_0^{T_1} \{f(X(s)) - r(f)\} ds$$

and (6.3). □

2.7 Efficiency of Regenerative Simulation

The results of Sections 2.6 and 2.7 can be used to assess the statistical efficiency of regenerative simulation when the regenerative process $\{X(t): t \geq 0\}$ is an irreducible CTMC with a finite state space, S. Take $S = \{0, 1, ..., N\}$ and observe that since $\{X(t): t \geq 0\}$ is necessarily positive recurrent, it is a regenerative process in continuous time and $X(t) \Rightarrow X$ as $t \to \infty$. Let f be a real-valued function having domain S and suppose that we wish to obtain estimates for the quantity

$$r(f) = \frac{E_0\{Y_1(f)\}}{E_0\{\tau_1(0)\}}.$$

from regenerative cycles formed by the successive entrances of the CTMC to state 0. (We assume that the regenerative process and the function f are such that the ratio formula for $r(f)$ holds.) As before, we denote by $T_n(0)$ the nth entrance time of $\{X(t):t \geq 0\}$ to state 0, $n \geq 1$; e.g.,

$$T_1(0) = \inf\{s > 0 : X(s-) \neq 0, X(s) = 0\}.$$

Set $T_0(0) = 0$ and $\tau_n(0) = T_n(0) - T_{n-1}(0)$, $n \geq 1$. For an assessment of the statistical efficiency of the simulation, it is convenient to have a c.l.t. in terms of simulation time, t.

Lemma 7.1. Suppose that $E_0\{\tau_1(0)\} < \infty$. Also suppose that $E_0\{Y_1(|f|)\} < \infty$ and $E_0\{Y_1^2(|f - r(f)|)\} < \infty$. Set $(\sigma_0(f))^2 = E_0\{Z_1^2(f)\}$. Then

(7.2)
$$\frac{t^{1/2}\left\{\int_0^t f(X(s))ds - r(f)t\right\}}{\sigma_0(f)/(E_0\{\tau_1(0)\})^{1/2}} \Rightarrow N(0,1)$$

as $t \to \infty$.

The half length of a confidence interval for $r(f)$ obtained from a simulation of fixed length, t, is proportional to the quantity appearing in the denominator of (7.2). Since the numerator in this c.l.t. is independent of the state 0 selected to form cycles, so is the denominator by the convergence of types theorem (Theorem 1.28 of Appendix 1). Thus the quantity $\sigma_0(f)/(E_0\{\tau_1(0)\})^{1/2}$ is an appropriate measure of the statistical efficiency of the simulation. The quantity can be obtained from the results of Section 2.6.

Example 7.3. (Cyclic Queues) Consider a queueing system consisting of two single-server service centers and a fixed number, N, of jobs. See Figure 2.1. After service completion at center 1, a job moves instantaneously to the tail of the queue at center 2, and after service completion at center 2 a job moves to the tail of the queue in center 1. Assume that both queues are served according to a FCFS discipline. Also suppose that all service times are mutually independent and that service times at center i identically distributed as a positive random variable, L_i, $i = 1,2$. Let $X(t)$ be the number of jobs waiting or in service at center 2 at time t and let event $e_i = $ "service completion at center i." The process $\{X(t):t \geq 0\}$ is a GSMP with finite state space, $S = \{0,1,...,N\}$ and event set, $E = \{e_1,e_2\}$.

If the service time random variable L_i is exponentially distributed (rate parameter λ_i), the process $\{X(t):t \geq 0\}$ is an irreducible, positive recurrent CTMC with finite state space. Moreover, the successive entrances to any fixed state $s \in S$ form a sequence of regeneration points for the process and $X(t) \Rightarrow X$ as $t \to \infty$. Let f be the indicator function, $1_{\{0,1,...,N-1\}}$, of the set $\{0,1,...,N-1\}$ so that $r(f)$ is the limiting probability that center 1 is busy. (Recall that for a set A, $1_A(x) = 1$ if $x \in A$ and equals 0 otherwise.) Tables 2.1 and 2.2 give theoretical values for estimation of $r(f)$ by regenerative simulation.

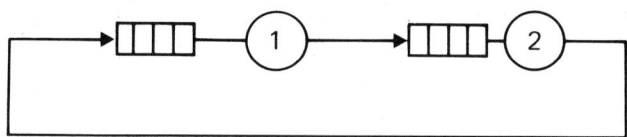

Figure 2.1. Cyclic queues

Table 2.1 Theoretical Values for Cyclic Queues
$N = 3$, $\lambda_1 = 4.0$, $\lambda_2 = 1.0$
Exponentially Distributed Service Times

	\multicolumn{4}{c}{Return State i}			
	0	1	2	3
$E_i\{Y_1(f)\}$	5.2500	1.0500	0.2625	0.3281
$E_i\{\tau_1(i)\}$	21.2500	4.2500	1.0625	1.3281
$E_i\{Y_1(f)\}/E_i\{\tau_1(i)\}$	0.2471	0.2471	0.2471	0.2471
$\sigma_i(f)$	1.5707	0.7024	0.3512	0.3927
$(\sigma_i(f))^2/E_i\{\tau_1(i)\}$	0.1161	0.1161	0.1161	0.1161

Table 2.2 Theoretical Values for Cyclic Queues
$N = 3$, $\lambda_1 = 1.25$, $\lambda_2 = 1.0$
Exponentially Distributed Service Times

	\multicolumn{4}{c}{Return State i}			
	2	1	2	3
$E_i\{Y_1(f)\}$	3.0500	1.3556	1.0844	1.9520
$E_i\{\tau_1(i)\}$	4.6125	2.0500	1.6400	2.9520
$E_i\{Y_1(f)\}/E_i\{\tau_1(i)\}$	0.6612	0.6612	0.6612	0.6612
$\sigma_i(f)$	1.4520	0.9680	0.8658	1.1616
$(\sigma_i(f))^2/E_i\{\tau_1(i)\}$	0.4571	0.4571	0.4571	0.4571

2.8 Regenerative Generalized Semi-Markov Processes

Although steady state estimation for an arbitrary GSMP is a formidable problem, estimation procedures are available for a GSMP that is a regenerative process in continuous time. To establish the regenerative property for a GSMP, it is necessary to show the existence of an infinite sequence of random time points at which the process probabilistically restarts.

Proposition 8.1 prescribes conditions on the building blocks of an irreducible, finite state GSMP with unit speeds that ensure that the process is a regenerative process in continuous time and that the expected time between regeneration points is finite. (Recall that a GSMP having state space, S, event set, E, and unit speeds is irreducible if for each pair $s,s' \in S$ there exists a finite sequence of states $s_1, s_2, \ldots, s_n \in S$ and events $e_{i_0}, e_{i_1}, \ldots, e_{i_n} \in E$ such that

$$p(s_1; s, e_{i_0}) p(s_2; s_1, e_{i_1}) \cdots p(s'; s_n, e_{i_n}) > 0.)$$

Proposition 8.1. Let $\{X(t): t \geq 0\}$ be an irreducible GSMP with a finite state space, S, event set, E, and unit speeds. Suppose that for all $s,s' \in S$, $e^* \in E$, and $e' \in N(s'; s, e^*)$ the clock setting distribution $F(\cdot; s', e', s, e^*)$ has a finite mean and a density function that is continuous and positive on $(0, +\infty)$. Also suppose that there exist $s_0, s_0' \in S$ and $e \in E$ such that for $e^* \in E(s_0)$,

(i)   the set $O(s_0'; s_0, e^*) = \emptyset$,
(ii)   the set $N(s_0'; s_0, e^*) = N(s_0'; s_0, e)$; and
(iii)   the clock setting distribution $F(\cdot; s_0', e', s_0, e^*) = F(\cdot; s_0', e', s_0, e)$ for all $e' \in N(s_0'; s_0, e^*)$.

Then $\{X(t): t \geq 0\}$ is a regenerative process in continuous time and the expected time between regeneration points is finite.

Under the conditions of Proposition 8.1, the process $\{X(t): t \geq 0\}$ makes a transition from state s_0 to state s_0' infinitely often with probability one. To see this, let ζ_n be nth time at which the process

$\{X(t): t \geq 0\}$ makes a state transition, $n \geq 0$. Also let $L(t)$ be the last state occupied by $\{X(t): t \geq 0\}$ before jumping to the state occupied at time t and set $V(t) = (L(t), X(t))$, $t \geq 0$. The process $\{V(t): t \geq 0\}$ is an irreducible GSMP with finite state space. Set $v_0' = (s_0, s_0')$ and associate the vector, $C(v_0')$, of clock readings with state v_0'. It is sufficient to show that the GSSMC associated with state transitions of the GSMP $\{V(t): t \geq 0\}$ returns infinitely often to the set $\{v_0'\} \times C(v_0')$; it follows immediately that $P\{V(\zeta_n) = v_0' \text{ i.o.}\} = 1$. The GSSMC returns infinitely often to the set $\{v_0'\} \times C(v_0')$ provided that:

(i) the GSMP is irreducible;
(ii) each clock setting distribution has finite mean and a density function that is continuous and positive on $(0, +\infty)$; and
(iii) a "recurrence measure" assigns positive measure to the set $\{v\} \times C(v)$ for all states v of the process.

Next observe that when $X(\zeta_n) = s_0'$ and $X(\zeta_n-) = s_0$ (i.e., $V(\zeta_n) = v_0'$), the only clocks that are active have just been set since $O(s_0'; s_0, e^*) = \emptyset$ for all $e^* \in E(s_0)$. The joint distribution of $X(\zeta_n)$ and the clocks set at time ζ_n depends on the past history of $\{X(t): t \geq 0\}$ only through s_0', the previous state, s_0, and the trigger event, e^*. As the new events and clock setting distributions are the same for all e^*, the process $\{X(t): t \geq 0\}$ probabilistically restarts whenever $\{X(\zeta_n): n \geq 0\}$ makes a transition from state s_0 to state s_0'. Since the state space, S, is finite and all clock setting distributions have finite mean, the expected time between regeneration points is finite.

Example 8.2 (Cyclic Queues) Suppose that all service times are mutually independent and that service times at center i have finite mean and a density function that is continuous and positive on $(0, +\infty)$, $i = 1, 2$. Let $X(t)$ be the number of jobs waiting or in service at center 1 at time t and let event $e_i = $ "service completion at center i." The process $\{X(t): t \geq 0\}$ is a GSMP with finite state space, $S = \{0,1,...,N\}$, event set, $E = \{e_1,e_2\}$, and unit speeds. Let ζ_n be the nth time at which the process $\{X(t):t \geq 0\}$ makes a state transition. Take $s_0' = 1$ and $s_0 = 0$ and observe that $X(\zeta_n) = 1$ and $X(\zeta_{n-1}) = 0$ only if the trigger event at time ζ_n is e_2. Also observe that $O(1;0,e_2) = \emptyset$ and $N(1;0,e_2) = \{e_1,e_2\}$. By Proposition 8.1, the successive times at which $\{X(\zeta_n):n \geq 0\}$ makes a transition from state s_0 to state s_0' are regeneration points for the process $\{X(t):t \geq 0\}$.

Lemma 8.3 is a special case of a generalized Borel-Cantelli lemma. The elementary proof given below uses a geometric trials argument. This result provides a means of showing that a GSMP with countable state space returns infinitely often (i.o.) to a fixed state with probability one.

Lemma 8.3. (Geometric Trials Lemma) Let $\{Y_n:n \geq 0\}$ be a sequence of random variables defined on a probability space $(\Omega,\mathcal{F},P)$ and taking values in a set, S. Let $s' \in S$. Suppose that there exists $\delta > 0$ such that

(8.4) $$P\{Y_n = s' \mid Y_{n-1},...,Y_0\} \geq \delta \text{ a.s..}$$

for all $n \geq 1$. Then $P\{Y_n = s' \text{ i.o.}\} = 1$.

Proof. Let I be the index of first entrance time of $\{Y_n:n \geq 0\}$ to state s':

$$I = \min\{n \geq 1: Y_n = s'\}.$$

Then

$$P\{I > n\} = P\{Y_n \neq s',...,Y_1 \neq s'\}$$

and it is sufficient to show that $P\{I > n\} \leq (1 - \delta)^n$ for all $n \geq 1$. For any n,

$$P\{I > n\} = P\{Y_n \neq s',...,Y_1 \neq s'\}$$

$$= E\{P\{Y_n \neq s',...,Y_1 \neq s' \mid Y_{n-1},...,Y_1\}\}$$

$$= E\{1_{\{Y_{n-1}\neq s',\ldots,Y_1\neq s'\}} P\{Y_n \neq s' \mid Y_{n-1},\ldots,Y_1\}\}$$

$$\leq E\{1_{\{Y_{n-1}\neq s',\ldots,Y_1\neq s'\}}(1-\delta)\}$$

$$= (1-\delta)\, P\{I > n-1\}$$

and therefore $P\{I > n\} \leq (1-\delta)^n$. □

Let $\{X(t): t \geq 0\}$ be a GSMP with countable state space, S, and event set, E. Suppose that $\{T_n : n \geq 0\}$ is an increasing sequence of finite ($T_n < \infty$ a.s.) state transition times such that for some $e^* \in E$ and $S^* \subseteq S$: $T_0 = 0$ and

(8.5) $\quad T_n = \inf\{t > T_{n-1}:$ at time t event e^* triggers

a transition in some state $s^* \in S^*\}$,

$n \geq 1$. Lemma 8.6 is an immediate consequence of Lemma 8.3 (with $Y_n = X(T_n)$). The geometric trials recurrence criterion in Lemma 8.6 avoids the "positive density on the half line" assumptions used in Proposition 8.1.

Lemma 8.6. Let $\{T_n : n \geq 0\}$ be an increasing sequence of finite ($T_n < \infty$ a.s.) state transition times as in (8.5). Let $s' \in S$ and suppose that

$$P\{X(T_n) = s' \mid X(T_{n-1}),\ldots,X(T_0)\} \geq \delta \text{ a.s.}$$

for some $\delta > 0$. Then $P\{X(T_n) = s' \text{ i.o.}\} = 1$.

Proposition 8.7 prescribes conditions that ensure that a GSMP with countable state space is a regenerative process in continuous time and the expected time between regeneration points is finite.

Proposition 8.7. Let $\{T_n : n \geq 0\}$ be an increasing sequence of stopping times that are finite ($T_n < \infty$ a.s.) state transition times as in (8.5). Suppose that there exist $s, s_0' \in S$ and $\delta > 0$ such that

(8.8) $\quad P\{X(T_n) = s_0' \mid X(T_{n-1}), \ldots, X(T_0)\} \geq \delta$ a.s.

and that for $s^* \in S^*$,

(i) the set $O(s_0'; s^*, e^*) = \emptyset$,
(ii) the set $N(s_0'; s^*, e^*) = N(s_0'; s, e^*)$; and
(iii) the clock setting distribution $F(\cdot; s_0', e', s^*, e^*) = F(\cdot; s_0', e', s, e^*)$ for all $e' \in N(s_0'; s^*, e^*)$.

Then $\{X(t): t \geq 0\}$ is a regenerative process in continuous time. Moreover, if

$$E\{T_{n+1} - T_n\} \leq c < \infty$$

for all $n \geq 0$, then the expected time between regeneration points is finite.

Proof. Using Lemma 8.6, (8.8) implies that event e^* triggers a transition to state s_0' infinitely often with probability one: $P\{X(T_n) = s_0' \text{ i.o.}\} = 1$. Furthermore, at such a time, T_n, the only clocks that are active have just been set since $O(s_0'; s^*, e^*) = \emptyset$ for all $s^* \in S$. The joint distribution of $X(T_n)$ and the clocks set at time T_n depends on the past history of $\{X(t): t \geq 0\}$ only through s_0', the previous state, s^*, and the trigger event, e^*. Since the new events and clock setting distributions are the same for all s^*, the process $\{X(t): t \geq 0\}$ probabilistically restarts whenever $\{X(T_n): n \geq 0\}$ hits state s_0'.

To show that the expected time between regeneration points is finite, assume for convenience that $X(T_0) = X(0) = s_0'$. Set $X_n = X(T_n)$ and $D_n = T_{n+1} - T_n$, $n \geq 0$. Observe that the random indices β_n such that $X_{\beta_n} = X(T_{\beta_n}) = s_0'$ form a sequence of regeneration points for the process $\{(X_n, D_n): n \geq 0\}$; this follows from that the fact the process $\{D_n: n \geq 1\}$ starts from scratch when $X(T_{\beta_n}) = s_0'$. Set $\alpha_k = \beta_{k+1} - \beta_k$, $k \geq 1$. The α_k are i.i.d. as α_1 and the argument in the proof of Lemma 8.3 shows that

2.8 Regenerative Generalized Semi-Markov Processes

$$P\{\alpha_1 > n\} \leq (1-\delta)^n$$

so that $E\{\alpha_1\} < \infty$ and the expected time between regeneration points for the process $\{(X_n, D_n) : n \geq 0\}$ is finite. Since $E\{\alpha_1\} < \infty$ and (8.8) ensures that α_1 is aperiodic, $(X_n, D_n) \Rightarrow (X, D)$ as $n \to \infty$. The continuous mapping theorem implies that $D_n \Rightarrow D$ as $n \to \infty$, and (since $D_n \geq 0$ and $E\{D_n\} \leq c < \infty$)

$$E\{D\} \leq \lim_{n \to \infty} E\{D_n\} \leq c < \infty.$$

Since α_1 is aperiodic, $E\{\alpha_1\} < \infty$, and $E\{|D|\} < \infty$,

$$E\{|D|\} = E\{D\} = \frac{E\left\{\sum_{j=0}^{\alpha_1-1} D_n\right\}}{E\{\alpha_1\}}$$

so that

$$E\left\{\sum_{j=0}^{\alpha_1-1} D_n\right\} < \infty$$

and the expected time between regeneration points for $\{X(t) : t \geq 0\}$ is finite. $\square$

The conclusion of Proposition 8.7 also holds if condition (i) is replaced by: (i') $O(s_0'; s^*, e^*) \neq \emptyset$ and for any $e' \in O(s_0'; s^*, e^*)$ the clock setting distribution $F(\cdot; s', e', s, e)$ is exponential with mean λ^{-1} independent of $s, s',$ and e. (Assumption (i') ensures that no matter when the clock for event $e' \in O(s_0'; s_0, e^*)$ was set, the remaining time until event e' triggers a state transition is exponentially distributed with mean λ^{-1}.) Note that the state transition times $\{T_n : n \geq 0\}$ are necessarily stopping times if

(8.9) $$p(s^*; s^*, e^*) = 0$$

for all $s^* \in S^*$ and

(8.10) $\quad e = e^*$ whenever $p(s;s^*,e) > 0$ and $p(s;s^*,e^*) > 0$

for all $s^* \in S^*$ and $s \in S$. (The conditions in (8.9) and (8.10) imply that every occurrence of event e^* in a state $s^* \in S^*$, and hence every state transition time T_n, can be determined by observing the sample paths of $\{X(t): t \geq 0\}$.)

Since the state space, S, of $\{X(t): t \geq 0\}$ is discrete and the expected time between regeneration points is finite, Proposition 2.3 implies that $X(t) \Rightarrow X$ as $t \to \infty$. Let f be a real-valued (measurable) function having domain S and set $r(f) = E\{f(X)\}$. From n cycles

$$(8.11) \qquad \hat{r}(n) = \frac{\bar{Y}(n)}{\bar{\tau}(n)} = \frac{\sum_{m=1}^{n} Y_m(f)}{\sum_{m=1}^{n} \alpha_m(f)}$$

is a strongly consistent point estimate for $r(f)$ and an asymptotic $100(1 - 2\gamma)\%$ confidence interval for $r(f)$ is

$$(8.12) \qquad \left[\hat{r}(n) - \frac{z_{1-\gamma} s(n)}{\bar{\tau}(n) n^{1/2}}, \hat{r}(n) + \frac{z_{1-\gamma} s(n)}{\bar{\tau}(n) n^{1/2}} \right].$$

The quantity $s^2(n)$ is a strongly consistent point estimate for $\sigma^2(f) = \text{var}(Y_1(f) - r(f)\tau_1)$. Asymptotic confidence intervals are based on the c.l.t.

$$(8.13) \qquad \frac{n^{1/2} \{\hat{r}(n) - r(f)\}}{\sigma(f)/E\{\tau_1\}} \Rightarrow N(0,1)$$

as $n \to \infty$.

The c.l.t. in (8.13) (and thus (8.12)) holds if $\sigma(f) < \infty$. It can be shown that if S is finite or f is bounded, $\sigma(f) < \infty$ provided that

$$E\{(T_{n+1} - T_n)^{2+\varepsilon}\} \leq b < \infty$$

for some $\varepsilon > 0$.

2.8 Regenerative Generalized Semi-Markov Processes 57

Lemma 8.6 can be used in conjunction with "new better than used" distributional assumptions to establish recurrence in a GSMP with countable state space.

Definition 8.14. The distribution F of a positive random variable A is *new better than used* (NBU) if

$$P\{A > x + y | A > y\} \leq P\{A > x\}$$

for all $x,y \geq 0$.

Note that every increasing failure rate (IFR) distribution is NBU. Also, if A and B are independent random variables with NBU distributions, then the distributions of $A + B$, $\min(A,B)$, and $\max(A,B)$ are NBU.

Example 8.15. (Cyclic Queues) Suppose that all service times are mutually independent and that service times at center i are identically distributed as a positive random variable, L_i, $i = 1,2$. Also suppose that L_2 has an NBU distribution and that $\delta = P\{L_2^{N-1} \leq L_1\} > 0$. Let $X(t)$ be the number of jobs waiting or in service at center 1 at time t. Take $e^* = e_1$ and $S^* = \{1,2,...,N\}$ so that T_n defined by (8.5) is the time of the nth service completion at center 1. (The state transition times $\{T_n : n \geq 0\}$ are stopping times since (8.9) and (8.10) hold.) Set $s_0' = N - 1$. It can be shown that

$$P\{X(T_n) = s_0' | X(T_{n-1}),...,X(T_0)\} \geq \delta \text{ a.s.}$$

so that $P\{X(T_n) = s_0' \text{ i.o.}\} = 1$ by Lemma 8.3. (The NBU assumption implies that the remaining center 2 service time at the start of a center 1 service time is stochastically dominated by an independent sample from the center 2 service time distribution.) Observe that $O(N - 1;N,e^*) = \emptyset$ and $N(N - 1;N,e_1) = \{e_1,e_2\}$ so that service times at center 1 and center 2 start afresh each time $\{X(T_n): n \geq 0\}$ hits state s_0'. By Proposition 8.7, the successive times T_n at which $X(T_n) = s_0'$ are regeneration points for the process $\{X(t): t \geq 0\}$.

Chapter 3

Markovian Networks of Queues

Networks of queues with priorities among job classes arise frequently as models for a wide variety of congestion phenomena. Simulation is usually the only available means for studying such networks. The underlying stochastic process of the simulation is defined in terms of a linear "job stack," an enumeration by service center and job class of all the jobs.

The job stack process for a closed network with Markovian job routing and exponential service times is a continuous time Markov chain with finite state space. Under the assumption that jobs queue at a center and receive service according to a fixed priority scheme among job classes, the job stack process need not be irreducible. (There may be one or more transient states and more than one irreducible, closed set of recurrent states.) Proposition 1.9 provides conditions on the building blocks of a network of queues that ensure that the set of recurrent states of the job stack process is irreducible. It follows that strongly consistent point estimates and asymptotic confidence intervals for general characteristics of the limiting distribution of the job stack process can be obtained by restricting the simulation to the set of recurrent states and applying the regenerative method of Chapter 2.

3.1 Markovian Job Stack Processes

We consider closed networks of queues having a finite number of *jobs* (customers) N, a finite number of *service centers*, s,

3.1 Markovian Job Stack Processes

and a finite number of (mutually exclusive) *job classes*, c. At every epoch of continuous time each job is of exactly one job class, but jobs may change class as they traverse the network. Upon completion of service at center i a job of class j goes to center k and changes to class l with probability $p_{ij,kl}$, where

$$P = \{p_{ij,kl} : (i,j),(k,l) \in C\}$$

is a given irreducible stochastic matrix and $C \subseteq \{1,2,...,s\} \times \{1,2,...,c\}$ is the set of (center, class) pairs in the network. In accordance with the matrix P, some centers may only see jobs of certain classes.

At each service center, jobs queue and receive service according to a fixed priority scheme among classes; the priority scheme may differ from center to center. Within a class at a center, jobs receive service according to a fixed queue service discipline. According to a fixed procedure for each center, a job in service may or may not be preempted if another job of higher priority joins the queue at the center. A job that has been preempted receives additional service at the center before any other job of its class at the center receives service.

We assume that all service times are mutually independent, and at a center have an exponential distribution with parameter that may depend on the service center, the class of job in service, and the "state" (as defined below) of the entire network when the service begins. In order to characterize the state of the network at time t, we let $S_i(t)$ be the class of the job receiving service at center i at time t, where $i = 1,2,...,s$; by convention $S_i(t) = 0$ if at time t there are no jobs at center i. If center i has more than one server, we take $S_i(t)$ to be a vector that records the class of the job receiving service from each server at center i. (Specifically, we enumerate the servers at center i as $1,2,...,s(i)$ and set

$$S_i(t) = (S_{i,1}(t), S_{i,2}(t),...,S_{i,s(i)}(t)),$$

where $S_{i,m}(t)$ is the class of the job receiving service from server m at center i at time t.) The classes of jobs serviced at center i in decreasing priority order are $j_1(i), j_2(i), \ldots, j_{k(i)}(i)$, elements of the set $\{1, 2, \ldots, c\}$. Let $C_{j_1}^{(i)}(t), \ldots, C_{j_{k(i)}}^{(i)}(t)$ be the number of jobs in queue at time t of the classes of jobs serviced at center i, $i = 1, 2, \ldots, s$. (This is shorthand notation: $C_{j_l}^{(i)}(t) = C_{j_l(i)}^{(i)}(t)$, $l = 1, 2, \ldots, k(i)$.)

We think of the N jobs being ordered in a linear stack (column vector) according to the following scheme. For $t \geq 0$ define the state of the system at time t to be the vector $Z(t)$ given by

(1.1) $Z(t) = \left(C_{j_{k(1)}}^{(1)}(t), \ldots, C_{j_1}^{(1)}(t), S_1(t); \ldots; C_{j_{k(s)}}^{(s)}(t), \ldots, C_{j_1}^{(s)}(t), S_s(t) \right).$

The *job stack at time t* then corresponds to the nonzero components in the vector $Z(t)$ and thus is an ordering of the jobs by class at the individual centers. Within a class at a particular service center, jobs waiting appear in the job stack in order of their arrival at the center, the latest to arrive being closest to the top of the stack. (A job that has been preempted appears at the head of its job class queue.) The process $Z = \{Z(t) : t \geq 0\}$ is called the *job stack process*.

For any service center i that sees only one job class (i.e., a center such that $k(i) = 1$) it is possible to simplify the state vector by replacing $C_{j_{k(i)}}^{(i)}(t), S_i(t)$ by $Q_i(t)$, the total number of jobs at center i. The state definition in (1.1) does not take into account explicitly that the total number of jobs in the network is fixed. For a complex network, use of this resulting somewhat larger state space facilitates generation of the process; for relatively simple networks, it may be desirable to remove the redundancy.

Example 1.2. (Cyclic Queues With Preemptive Priority) Consider a network with two single server service centers and two job classes. Suppose that the set, C, of (center, class) pairs is $C = \{(1,1), (2,1), (2,2)\}$ (so that center 1 serves jobs of class 1 and center 2 serves jobs of class 1 and class 2), all queueing disciplines are FCFS, and jobs of class 2 have preemptive priority at center 2 over jobs of class 1. Also suppose that a job of class 1 completing service at center 2 joins the tail of the queue at center 1 with probability p and with probability $1 - p$ joins the tail of the class 2 queue at center 2. A job of class 2 completing service at center 2 joins the tail of the queue at center 1. A job completing service at center 1 joins the tail of the class 1 queue at center 2. Thus,

$$P = \begin{matrix} 0 & 1 & 0 \\ p & 0 & 1-p \\ 1 & 0 & 0 \end{matrix}.$$

Set

(1.3) $$Z(t) = \bigl(Q_1(t), C_1^{(2)}(t), C_2^{(2)}(t), S_2(t)\bigr),$$

where $Q_1(t)$ is the number of jobs waiting or in service at center 1 at time t, $S_2(t)$ is the class of the job in service at center 2, and $C_j^{(2)}(t)$ is the number of jobs of class j waiting in queue at center 2, $j = 1,2$. (Observe that $j_1(1) = 1$, $j_1(2) = 2$, and $j_2(2) = 1$.) When there are $N = 2$ jobs in the network the state space, D^*, of the job stack process $Z = \{Z(t): t \geq 0\}$ is

$$D^* = \{(2,0,0,0),(1,0,0,1),(0,1,0,1),(0,1,0,2),(1,0,0,2),(0,0,1,2)\}.$$

Since jobs of class 2 have preemptive priority over jobs of class 1 at center 2, $(0,0,1,1)$ is not a stable state. With $N = 5$ jobs, $z = (3,1,0,2)$ is a state of the job stack process. The job in position 1 of the job stack associated with state z is at the tail of the queue at center 1; the job in position 4 of the job stack is waiting in queue at center 2 as a job of class 1.

Proposition 1.4 is a direct consequence of the assumptions of Markovian job routing and exponential service times.

Proposition 1.4. The job stack process $Z = \{Z(t): t \geq 0\}$ is a continuous time Markov chain with finite state space, D^*.

The characterization of a CTMC in (1.5) below leads to an efficient procedure for generating sample paths of a Markovian job stack process. Let $X = \{X(t): t \geq 0\}$ be a (time-homogeneous) CTMC with finite state space, S, so that for any $t,s \geq 0$ and $j \in S$,

$$P\{X(t+s) = j \mid X(u); u \leq s\} = P\{X(t+s) = j \mid X(s)\}$$

and the conditional probability

$$p_{ij}(t) = P\{X(t+s) = j \mid X(s) = i\}$$

is independent of $t \geq 0$ for all $i,j \in S$ and $s \geq 0$. The CTMC is specified by an infinitesimal generator, $Q = (q_{ij})$, such that $q_{ij} \geq 0$ for $i,j \in S$ with $i \neq j$,

$$\sum_{j \in S} q_{ij} = 0$$

for all i, and $0 < q_i = -q_{ii} < \infty$. Denote the increasing sequence of jump times for the process by $\{\zeta_n : n \geq 0\}$ and set $X_n = X(\zeta_n)$, $n = 0, 1, \ldots$.

For $i \in S$, set $q_i = -q_{ii}$ and

$$r_{ij} = \begin{matrix} q_{ij}/q_i & \text{if } i \neq j \\ 0 & \text{if } i = j \end{matrix}$$

Then for any $j \in S$, $u > 0$, and $n = 0, 1, \ldots$,

(1.5) $\quad P\{X_{n+1} = j, \zeta_{n+1} - \zeta_n > u \mid X_0, \ldots, X_n; \zeta_0, \ldots, \zeta_n\}$

$\quad = r_{ij} \exp(-q_i u)$ for $X_n = i$.

According to (1.5), given a jump to state i, the CTMC X remains in state i for an exponentially distributed (mean q_i^{-1}) amount of time and then jumps to state j with independent probability r_{ij}. This means that a state transition for the CTMC can be generated by producing a pair of independent random numbers.

This pair consists of an exponential random number and a sample from a discrete distribution specified by the jump probabilities. The characterization of (1.5) can be used without explicit enumeration of the state space of the process and computation of the elements of the infinitesimal generator, Q. The routing matrix P and the rate parameters of the exponential service time distributions completely determine the infinitesimal generator of the job stack process for a network of queues.

Example 1.6. (Cyclic Queues) Assume that service times at center i are exponentially distributed random variables, L_i, with mean λ_i^{-1}, $i = 1,2$. Let $X(t)$ be the number of jobs waiting or in service at center 2 at time t. The process $\{X(t): t \geq 0\}$ is an irreducible CTMC with finite state space, $S = \{0,1,...,N\}$. The nonzero elements of the jump matrix R are $r_{01} = 1$, $r_{NN-1} = 1$, and

$$r_{ii+1} = 1 - r_{ii-1} = \lambda_1/(\lambda_1 + \lambda_2),$$

$i = 1,2,...,N-1$. The elements of the vector $q = (q_0,...,q_N)$ are $q_0 = \lambda_1$, $q_N = \lambda_2$, and $q_i = \lambda_1 + \lambda_2$, $i = 1,2,...,N-1$.

To see this, observe that in state 0 all N jobs are at center 1 and after an exponentially distributed (mean λ_1^{-1}) amount of time, there is a transition to state 1. Similarly, in state N all jobs are at center 2 and the next transition is to state $N-1$ after an exponential distributed (mean λ_2^{-1}) amount of time. In state i there is a job in service at center 1 and a job in service at center 2. A transition from state i to state $i+1$ corresponds to completion of service at center 1 (prior to completion at center 2), $1 \leq i \leq N-1$. Then

$$r_{ii+1} = 1 - r_{ii-1} = P\{L_1 \leq L_2\} = \lambda_1/(\lambda_1 + \lambda_2).$$

The holding time in state i is distributed as min (L_1, L_2). Since

$$\min(L_1, L_2) > t \text{ if and only if } L_1 > t \text{ and } L_2 > t$$

for any fixed $t > 0$, the holding time in state i is exponentially distributed with mean $(\lambda_1 + \lambda_2)^{-1}$.

Some restrictions on the building blocks of a network of queues with priorities among job classes are needed to ensure that a Markovian job stack process has a limiting distribution, independent of the initial state. This is the case if job stack process has a single irreducible, closed set of recurrent states. (By definition, the recurrent states of a CTMC are the recurrent states of its embedded jump chain. The recurrent states of any DTMC can be divided in a unique manner into irreducible, closed sets.)

Irreducibility of the routing matrix P does not ensure either that all states of the job stack process are recurrent or that there is a single irreducible, closed set of recurrent states. (For the network of Example 1.2 with $N = 2$ jobs, state $(0,0,1,2)$ of the job stack process is transient. The remaining five states are recurrent, and the set of recurrent states is irreducible. See Figure 3.1.) Example 1.7 shows that the job stack process can have more than one irreducible, closed set of recurrent states.

Example 1.7. Consider a network with two single server service centers and three job classes and suppose that the set, C, of (center, class) pairs is $C = \{(1,1),(1,2),(1,3),(2,1),(2,2),(2,3)\}$. Also suppose that all queueing disciplines are FCFS and that at each of the centers jobs of class 1 have nonpreemptive priority over jobs of class 2 and jobs of class 2 have nonpreemptive priority over jobs of class 3. Assume that upon completion of service at center 1, with probability one a job of class 1 goes to center 2. The job becomes class 2 at center 2 if it was class 1 at center 1 and becomes class 3 if it was class 2 at center 1; otherwise the job becomes class 1 at center 2. Upon completion of service at center 2, with probability one a job goes to center 1 but does not change class. See Figure 3.2.

3.1 Markovian Job Stack Processes

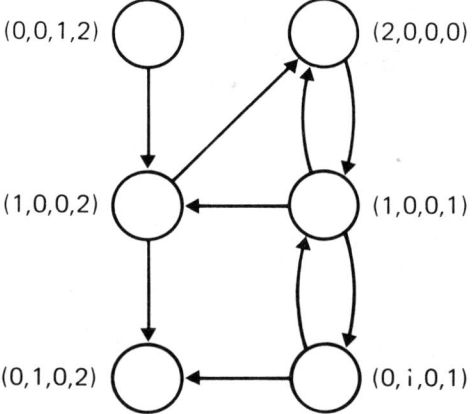

Figure 3.1. States of the job stack process

Set

$$Z(t) = \left(C_3^{(1)}(t), C_2^{(1)}(t), C_1^{(1)}(t), S_1(t), C_3^{(2)}(t), C_2^{(2)}(t), C_1^{(2)}(t), S_2(t)\right),$$

where $S_i(t)$ is the class of the job in service at center i at time t and $C_j^{(i)}(t)$ is the number of jobs of class j waiting in queue at center i, $i = 1,2$. With $N = 2$ jobs in the network all twenty seven states of the job stack process are recurrent, but there are two irreducible, closed sets of recurrent states. One set consists of nine states: (0,1,0,3,0,0,0,0), (0,0,0,0,0,1,0,3), (0,0,1,2,0,0,0,0), (0,0,0,0,0,0,1,2), (1,0,0,1,0,0,0,0), (0,0,0,0,1,0,0,1), (0,0,0,1,0,0,0,3), (0,0,0,3,0,0,0,2), and (0,0,0,2,0,0,0,1).

66 3 Markovian Networks

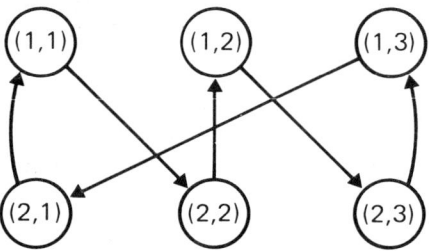

Figure 3.2. Markovian routing of (center, class) pairs

Proposition 1.9 asserts that the job stack process has a single irreducible, closed set of recurrent states provided that the routing matrix P is irreducible and there is a service center, i_0, that sees only one job class or is such that jobs of the lowest priority job class at the center are subject to preemption. The idea is to show that state z^* in which all N jobs are of (the lowest priority) class $j_{k(i_0)}(i_0)$ at center i_0 is a *target state* of the job stack process in the sense that it is accessible from any other state, z, of the embedded jump chain $\{Z_k : k \geq 0\}$. The proof is based on the existence of a finite length "path" from $(i_0, j_{k(i_0)}(i_0))$ to $(i_0, j_{k(i_0)}(i_0))$ that includes all (center, class) pairs. This implies the existence of a finite sequence of state transitions in $\{Z_k : k \geq 0\}$ such that the distance from z to z^* (as defined below) decreases to zero. For $z, z' \in D^*$ we write $z \to z'$ when the probability of transition from z to z' in one step of the embedded jump chain $\{Z_k : k \geq 0\}$ is positive; we write $z \sim z'$

when z' accessible from z: starting from z the probability of entering z' on the nth step is positive for some $n \geq 1$.

Lemma 1.8. Suppose that there exists a state, z^*, of the job stack process such that $z \sim z^*$ for all $z \in D^*$. Then the set, D, of recurrent states of the embedded jump chain $\{Z_k : k \geq 0\}$ is irreducible.

Proof. The set D is nonempty since D^* is finite. Assume that D is not irreducible. Then there exist $z_1, z_2 \in D$ such that z_1 is not accessible from z_2. Let $R_1 = \{z \in D^* : z_1 \sim z\}$ be the set of all states of $\{Z_k : k \geq 0\}$ that are accessible from z_1 and observe that $z^* \in R_1$. As z_1 is recurrent, the set R_1 is irreducible and therefore $z^* \sim z_1$. But $z_2 \sim z^*$ by hypothesis and thus $z_2 \sim z_1$, a contradiction. □

Proposition 1.9. Suppose that the routing matrix P is irreducible and that for some service center, i_0, either $k(i_0) = 1$ or service to a job of class $j_{k(i_0)}(i_0)$ at center i_0 is preempted when any other job of higher priority joins the queue. Then the set, D, of recurrent states of the embedded jump chain $\{Z_k : k \geq 0\}$ is irreducible.

Proof. By Lemma 1.8, it is sufficient to show the existence of a state, z^*, of the job stack process such that $z \sim z^*$ for all $z \in D^*$. Without loss of generality, assume that $i_0 = 1$: either $k(1) = 1$ or service to jobs of class $j_{k(1)}(1)$ at center 1 is subject to preemption when jobs of higher priority join the queue. Let z^* be the state in which there is one job of class $j_{k(1)}(1)$ in service at center 1 and $N - 1$ jobs of class $j_{k(1)}(1)$ in queue at center 1 (or in service if center 1 is a multiple server center). As the routing matrix P is irreducible, there exists a finite sequence of (center, class) pairs $(i_1, j_1), (i_2, j_2), \ldots, (i_M, j_M) \in C$ such that (i) $(i_1, j_1) = (i_M, j_M) = (1, j_{k(1)}(1))$, (ii) for any (center, class) pair $(i, j) \in C$ there exists n ($1 \leq n \leq M$) such that $(i_n, j_n) = (i, j)$, and (iii) $p_{i_m j_m, i_{m+1} j_{m+1}} > 0$ for $m = 1, 2, \ldots, M - 1$. Let l_m be the index of the first occurrence of

$(1, j_{k(1)}(1))$ following (i_m, j_m):

$$l_m = \min \{l \geq m : (i_l, j_l) = (1, j_{k(1)}(1))\},$$

$m = 1, 2, \ldots, M - 1$. Fix $(i, j) \in C$. Then there exists a subsequence $(i_m, j_m), (i_{m+1}, j_{m+1}), \ldots, (i_{l_m}, j_{l_m})$ of $(i_1, j_1), (i_2, j_2), \ldots, (i_M, j_M)$ such that $(i_m, j_m) = (i, j)$. Select the shortest such subsequence. (If there are several of equal length, shorter than all other, select the one for which the index of the first element is smallest.) Let $m(i, j)$ be the index of the first element of this subsequence:

$$m(i, j) = \min \{m : (i_m, j_m) = (i, j) \text{ and } l_m - m \leq l_n - n$$

$$\text{for all } n \text{ such that } (i_n, j_n) = (i, j)\}.$$

In terms of this index define the successor (center, class) pair, $s(i, j)$, of (i, j) as $s(i, j) = (i_{m(i,j)+1}, j_{m(i,j)+1})$.

For $z \in D^*$, let $U(z)$ be the set of all (center, class) pairs $(i, j) \in C - \{(1, j_{k(1)}(1))\}$ such that when the job stack process is in state z there is at least one job of class j in service at center i. Let h be a function taking values in C and having domain $D^* \times \{1, 2, \ldots, N\}$ such that $h(z, n)$ is (i, j) when the job in position n of the job stack associated with state z is of class j at center i. Define a nonnegative distance from state z to state z^* as follows. For the job in position n of the job stack associated with state z, set

$$d(z, n; z^*) = \min \{l_m - m : (i_m, j_m) = h(z, n)\},$$

$n = 1, 2, \ldots, N$. Then define the distance, $d(z; z^*)$, from z to z^* as

$$d(z, z^*) = \sum_{n=1}^{N} d(z, n; z^*).$$

First suppose that $z \neq z^*$. Then $U(z)$ is nonempty since either $k(1) = 1$ or service to jobs of class $j_{k(1)}(1)$ at center 1 is subject to preemption. Select $(k, l) \in U(z)$ and let z_1 be the neighbor of z having one more job of class $j_{m(k,l)+1}$ at center $i_{m(k,l)+1}$ and one less job of class l at center k. As $(k, l) \in U(z)$, it follows from the definition of the successor (center, class) pair $s(k,l)$ of (k,l) that $z \to z_1$. Moreover, $d(z_1;z^*) < d(z;z^*)$. Next, select $(k_1,l_1) \in U(z_1)$ and let z_2 be the neighbor of z_1 having one more job of (center, class) pair $s(k_1,l_1)$ and one less job of class l_1 at center k_1. Clearly, $z_1 \to z_2$ and necessarily $d(z_2;z^*) < d(z_1;z^*)$. Continuing in this way for at most a finite number of steps, the distance to z^* decreases to zero and it follows that $z \sim z^*$. Now suppose that $z = z^*$. Then there exists $z' \neq z^*$ such that $z^* \to z'$, and by the previous argument $z \sim z^*$. □

A similar argument shows that $\{Z_k : k \geq 0\}$ has a single irreducible, closed set of recurrent states if the routing matrix P is irreducible and some service center sees only two job classes.

Example 1.10. (Cyclic Queues With Preemptive Priority) Set $Z(t) = \left(Q_1(t), C_1^{(2)}(t), C_2^{(2)}(t), S_2(t)\right)$ as in (1.3). Center 1 sees only one job class and $z^* = z_1^* = (N,0,0,0)$ is a target state of the job stack process in the sense that $z \sim z^*$ for all $z \in D^*$. Take $(i_1,j_1) = (1,1)$, $(i_2,j_2) = (2,1)$, $(i_3,j_3) = (2,2)$, and $(i_4,j_4) = (1,1)$ so that (i_1,j_1), (i_2,j_2), (i_3,j_3), (i_4,j_4) is a path of (center, class) pairs from (1,1) to (1,1) that includes all (center, class) pairs. Suppose that there are $N = 4$ jobs and take $z = (1,2,0,2)$. Set $z_1 = (2,1,0,1)$, $z_2 = (2,1,0,2)$, $z_3 = (3,0,0,1)$, and $z_4 = (3,0,0,2)$. Then $z \to z_1$, $z_1 \to z_2$, $z_2 \to z_3$, $z_3 \to z_4$, and $z_4 \to z^*$. See Figure 3.3. The symbol ● denotes a job in service and ○ denotes a job waiting in queue. Since jobs of class 2 have preemptive priority over jobs of class 1 at center 2, $z_2^* = (0,0,N-1,1)$ is also a target state of the job stack process.

Corollary 1.11 follows directly from Proposition 1.9.

Corollary 1.11. Suppose that the routing matrix P is irreducible and that for some service center, i_0, either $k(i_0) = 1$ or service to a job of class $j_{k(i_0)}(i_0)$ at center i_0 is preempted when any other job of higher priority joins the queue. Then restricted to the set, D, of recurrent states, the job stack process $Z = \{Z(t) : t \geq 0\}$ is irreducible and positive recurrent.

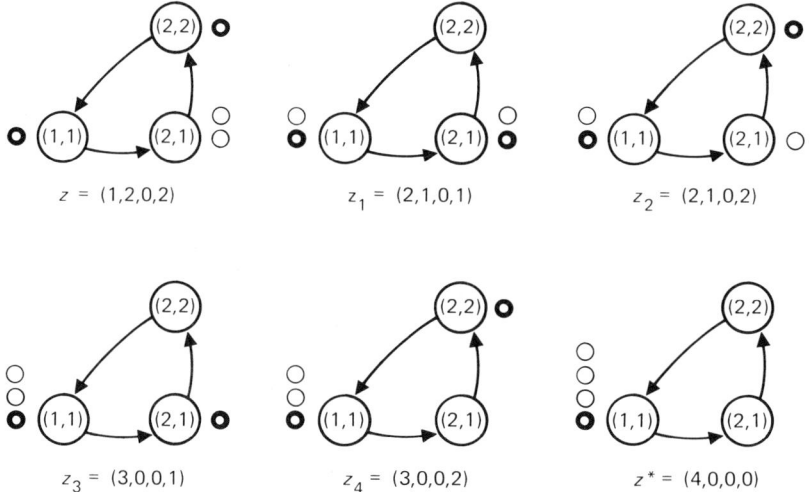

Figure 3.3. Target state of the job stack process

Example 1.12. (Data Base Management System Model) Consider (as a model of resource contention in a data base management system) a network of queues with two service centers that provides service to N (≥ 2) jobs; see Figure 3.4. Center 2 only serves jobs of class 1, and the classes of jobs served at center 1 in order of decreasing priority are $j_1(1) = 2$, $j_2(1) = 3$,..., $j_6(1) = 7$. Thus, the set, C, of (center, class) pairs in the network is

$$C = \{(1,2),(1,3),(1,4),(1,5),(1,6),(1,7),(2,1)\}.$$

Service to jobs of class 7 (at center 1) is subject to preemption when any other job of higher priority joins the queue at center 1.

3.1 *Markovian Job Stack Processes* 71

Service to jobs of any other job class is not subject to preemption. Suppose that for fixed $0 < p_1, p_2 < 1$ the routing matrix P is

$$P = \begin{pmatrix} 0 & 1-p_1 & 0 & 0 & 0 & 0 & p_1 \\ 1-p_2 & 0 & 0 & 0 & 0 & p_2 & 0 \\ 0 & 1-p_1 & 0 & 0 & 0 & 0 & p_1 \\ 1-p_2 & 0 & 0 & 0 & 0 & p_2 & 0 \\ 0 & 0 & 1 & 0 & 0 & 0 & 0 \\ 0 & 0 & 0 & 0 & 1 & 0 & 0 \\ 0 & 0 & 0 & 1 & 0 & 0 & 0 \end{pmatrix}.$$

Thus, for example, upon completion of service at center 1 a job of class 2 joins the tail of the queue at center 2 (as class 1) with probability p_1 and with probability $1 - p_1$ changes to class 3 and joins the tail of the queue at center 1.

Set

(1.13) $\quad Z(t) = \left(C_7^{(1)}(t), \ldots, C_2^{(1)}(t), S_1(t), Q_2(t)\right),$

where $C_j^{(1)}(t)$ is the number of jobs of class j in queue at center 1 at time t, $S_1(t)$ is the class of job in service at center 1, and $Q_2(t)$ is the number of jobs waiting or in service at center 2. Since $k(2) = 1$ and service to jobs of class $j_{k(1)}(1) = 7$ at center 1 is subject to preemption, either state $z_1^* = (N - 1, 0, 0, 0, 0, 0, 7, 0)$ or state $z_2^* = (0, 0, 0, 0, 0, 0, 0, N)$ can serve as the target state, $z_{i_0}^*$, for the job stack process. With $N \geq 2$ jobs, the set $D^* - D$ is nonempty; e.g., state $(0, 0, 0, k - 1, 0, 0, 4, N - k) \in D^* - D$ provided that $k \geq 2$.

Proposition 1.14 is a direct consequence of Corollary 1.11 and the definition of a regenerative process. The successive entrances to any fixed recurrent state form a sequence of regeneration points for the job stack process.

Proposition 1.14. Under the conditions of Proposition 1.9, the job stack process $Z = \{Z(t) : t \geq 0\}$ is a regenerative process in continuous time and the expected time between regeneration points is finite.

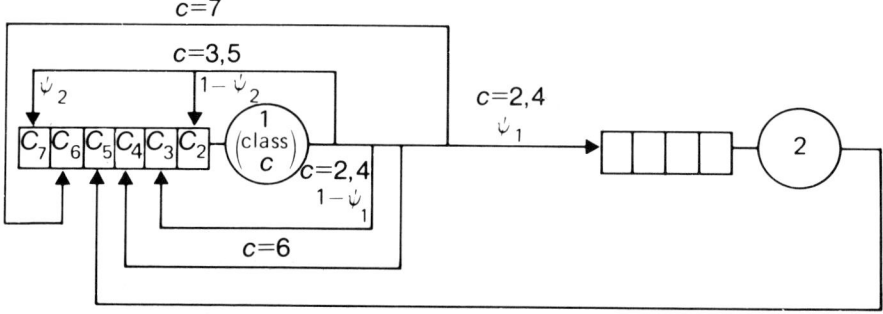

Figure 3.4. Data base management system model

Since the state space of the job stack process is discrete and the expected time between regeneration points is finite, $Z(t) \Rightarrow Z$ as $t \to \infty$ by Proposition 2.3 of Chapter 2. Let f be a real-valued function with domain, D, and suppose that the goal of the simulation is the estimation of $r(f) = E\{f(Z)\}$. Strongly consistent point estimates and asymptotic confidence intervals for $r(f)$ can be obtained by the standard regenerative method of Chapter 2. To obtain estimates for $r(f)$, select a recurrent state, z, of the job stack process Z and take $Z(0) = z$. Then set $T_0 = 0$ and

$$T_k = \inf\{t > T_{k-1} : Z(t) = z, Z(t-) \neq z\}$$

so that T_k is the kth time at which the process Z makes a transition to state z, $k \geq 1$. Propositions 1.15 and 1.16 follow from the general results for regenerative processes in Chapter 2. Set

$$Y_k(f) = \int_{T_{k-1}}^{T_k} f(Z(u))\,du$$

and $\tau_k = T_k - T_{k-1}$.

Proposition 1.15. The sequence of pairs of random variables $\{(Y_k(f), \tau_k): k \geq 1\}$ are independent and identically distributed.

Proposition 1.16. Provided that $E\{|f(Z)|\} < \infty$,

$$r(f) = \frac{E\{Y_1(f)\}}{E\{\tau_1\}}.$$

With these results the standard regenerative method applies. Based on n cycles

$$\hat{r}(n) = \frac{\bar{Y}(n)}{\bar{\tau}(n)} = \frac{\sum_{m=1}^{n} Y_m(f)}{\sum_{m=1}^{n} \tau_m}$$

is a strongly consistent point estimate for $r(f)$ and an asymptotic $100(1 - 2\gamma)\%$ confidence interval is

$$\left[\hat{r}(n) - \frac{z_{1-\gamma}\,s(n)}{\bar{\tau}(n)\,n^{1/2}},\, \hat{r}(n) + \frac{z_{1-\gamma}\,s(n)}{\bar{\tau}(n)\,n^{1/2}}\right].$$

The quantity $s^2(n)$ is a strongly consistent point estimate for $\sigma^2 = \text{var}\,(Y_1(f) - r(f)\tau_1)$. Using Wald's second moment identity and the fact that the service time distributions have finite second moment, it can be shown that the time between regeneration points has finite second moment. Since the state space of the job stack process is finite, it follows that the variance constant $\sigma^2 < \infty$. Asymptotic confidence intervals for $r(f)$ are based on the c.l.t.

$$\frac{n^{1/2}\{\hat{r}(n) - r(f)\}}{\sigma/E\{\tau_1\}} \Rightarrow N(0,1)$$

as $n \to \infty$.

Algorithm 1.17. (Regenerative Method for Markovian Networks)

1. Select a recurrent state, z, of the job stack process.
2. Set $Z(0) = z$ and simulate the job stack process. Observe a fixed number, n, of cycles defined by the successive times at which the process makes a transition to state z.
3. Compute the length, τ_m, of the mth cycle and the quantity

$$Y_m(f) = \int_{T_{m-1}}^{T_m} f(Z(u))du,$$

where $T_0 = 0$ and $T_m = \tau_1 + \ldots + \tau_m$.

4. Form the point estimate

$$\hat{r}(n) = \frac{\bar{Y}(n)}{\bar{\tau}(n)}.$$

5. Form the asymptotic $100(1 - 2\gamma)\%$ confidence interval

$$\left[\hat{r}(n) - \frac{z_{1-\gamma}\, s(n)}{\bar{\tau}(n)\, n^{1/2}}, \hat{r}(n) + \frac{z_{1-\gamma}\, s(n)}{\bar{\tau}(n)\, n^{1/2}}\right].$$

To obtain estimates for $r(f)$, we must select a recurrent state of the job stack process to serve as a return state. For complex networks of queues it is nontrivial to determine by inspection whether or not a state of the job stack process is recurrent.

Proposition 1.18 characterizes the set of recurrent states of a Markovian job stack process. The proof uses the fact that only recurrent states are accessible from a recurrent state of a DTMC.

Proposition 1.18. Suppose that the routing matrix P is irreducible. Also suppose that for some service center, i_0, either $k(i_0) = 1$ or service to a job of class $j_{k(i_0)}(i_0)$ at center i_0 is preempted when any other job of higher priority joins the queue. Let z_i^* be the state of the job stack process in which all N jobs are of class $j_{k(i)}(i)$ at center i, $i = 1,2,\ldots,s$. Then z is a recurrent state of the job stack process $Z = \{Z(t): t \geq 0\}$ if and only if $z_i^* \sim z$ for some $i \in \{1,2,\ldots,s\}$.

Proof. Without loss of generality, assume that $i_0 = 1$ so that either $k(1) = 1$ or service to a job of class $j_{k(1)}(1)$ at center 1 is preempted when any other job of higher priority joins the queue. First observe that each of the states z_i^* is recurrent, $i = 1,2,\ldots,s$. (By the argument in the proof of Proposition 1.9, $z \sim z_1^*$ for all $z \in D^*$. This implies that z_1^* is recurrent because a finite state DTMC has at least one recurrent state and only recurrent states are accessible from a recurrent state. By hypothesis, either $k(1) = 1$ or service to jobs of class $j_{k(1)}(1)$ at center 1 is preempted when any other job of higher priority joins the queue. In either case, it is easy to show that state z_i^* is accessible from state z_1^*, $i = 2,3,\ldots,s$. This implies that z_i^* is recurrent.) Therefore, state z is recurrent if $z_i^* \sim z$ for some i. Conversely the set of recurrent states of the embedded jump chain $\{Z_k: k \geq 0\}$ is irreducible by Proposition 1.9. Therefore, $z_i^* \sim z$ for all i if state z is recurrent. □

Proposition 1.18 ensures that a state z of the job stack process Z is recurrent if it is accessible from some state z_i^*. If state z is recurrent, it is accessible from all of the z_i^* by Proposition 1.9.

Example 1.19. (Data Base Management System Model) State $z = (N - 1,0,0,0,0,0,5,0)$ of the job stack process Z defined by (1.13) is recurrent since it is accessible from state $z_1^* =$

$(N-1,0,0,0,0,0,7,0)$. To see this, set $z_1 = (N-1,0,0,0,0,0,6,0)$, $z_2 = (N-1,0,0,0,0,0,4,0)$, and $z_3 = (N-2,0,0,0,0,0,7,1)$ and observe that $z_1^* \to z_1$, $z_1 \to z_2$, $z_2 \to z_3$, and $z_3 \to z$. A transition from state z_1^* to state z_1 occurs (with probability one) upon completion of service to a job of class 7 at center 1. With probability one a transition from state z_1 to state z_2 occurs upon completion of service to a job of class 6 at center 1. Upon completion of service to a job of class 4 at center 1, the process makes a transition from state z_2 to state z_3. Then, upon completion of service at center 1 and with preemption of the class 7 job in service at center 1, the process makes a transition from state z_3 to state z.

3.2 Augmented Job Stack Processes

Informally, passage times in a network of queues are the random times for a job to traverse a portion of the network. In order to obtain point and interval estimates for general characteristics of passage times, we augment the job stack used in Section 3.1 to describe the state of the network and measure individual passage times. A minimal state vector augmentation rests on the notion of an arbitrarily chosen, distinguished job. The idea is to keep track of the position in the job stack of the distinguished job and to measure passage times for this "marked job." Underlying regenerative process structure provides a ratio formula and a central limit theorem. These lead to strongly consistent point estimates and asymptotic confidence intervals.

Proposition 1.9 provides conditions on the building blocks of a network of queues that ensure that the set of recurrent states of the job stack process is irreducible. It follows that the job stack process has a limiting distribution independent of the initial state. These conditions imply that the augmented job stack process of a network with at least two service centers has a single irreducible, closed set of recurrent states. The "marked job method" developed in Section 3.4 provides point and interval estimates for general characteristics of limiting passage times. Estimates are obtained by simulating the augmented job stack process in random blocks defined by passage times for the marked job that start in a fixed state. The "labelled jobs" method of Section 3.6 provides point and interval estimates for passage times that correspond to passage through a subnetwork of a given network of queues. Observed passage times for all the jobs enter into the estimates.

Let $N(t)$ be the position (from the top) of the marked job in the job stack at time t. Then set

(2.1) $$X(t) = (Z(t), N(t)).$$

The process $X = \{X(t) : t \geq 0\}$ is called the *augmented job stack process*. Proposition 2.2 is a consequence of the assumptions of Markovian job routing and exponential service times.

Proposition 2.2. The augmented job stack process $X = \{X(t) : t \geq 0\}$ is a continuous time Markov chain with finite state space, G^*.

Passage times for the marked job are specified by means of four nonempty subsets $(A_1, A_2, B_1, \text{ and } B_2)$ of the state space, G^*, of the augmented job stack process X. The sets A_1, A_2 [resp., B_1, B_2] jointly define the random times at which passage times for the marked job start [resp., terminate]. The sets $A_1, A_2, B_1,$ and B_2 in effect determine when to start and stop the clock measuring a particular passage time of the marked job.

Denoting the increasing sequence of jump times of the augmented job stack process X by $\{\zeta_n : n \geq 0\}$, for $k, n \geq 1$ we require that the sets $A_1, A_2, B_1,$ and B_2 satisfy:

if $X(\zeta_{n-1}) \in A_1$, $X(\zeta_n) \in A_2$, $X(\zeta_{n-1+k}) \in A_1$, and $X(\zeta_{n+k}) \in A_2$
then $X(\zeta_{n-1+m}) \in B_1$ and $X(\zeta_{n+m}) \in B_2$ for some $0 < m \leq k$;

and

if $X(\zeta_{n-1}) \in B_1$, $X(\zeta_n) \in B_2$, $X(\zeta_{n-1+k}) \in B_1$, and $X(\zeta_{n+k}) \in B_2$
then $X(\zeta_{n-1+m}) \in A_1$ and $X(\zeta_{n+m}) \in A_2$ for some $0 \leq m < k$.

These conditions ensure that the start times and termination times for the specified passage time strictly alternate. (We assume that for all $x \in A_2$ there exists $x \in A_1$ such that $x_1 \to x_2$ and that for all $x_2 \in B_2$ there exists $x_1 \in B_1$ such that $x_1 \to x_2$. We write $x \to x'$ when there is a positive probability that the embedded jump chain $\{X(\zeta_n) : n \geq 0\}$ makes a transition in one step from state x to state x'.)

In terms of the sets A_1, A_2, B_1, and B_2, we define two sequences of random times, $\{S_j : j \geq 0\}$ and $\{T_j : j \geq 1\}$: S_{j-1} is the start time for the jth passage time for the marked job and T_j is the termination time of this jth passage time. Assume that the initial state of the augmented job stack process X is such that a passage time for the marked job begins at $t = 0$. Set $S_0 = 0$,

$$S_j = \inf \{\zeta_n \geq T_j : X(\zeta_n) \in A_2, X(\zeta_{n-1}) \in A_1\},$$

and

$$T_j = \inf \{\zeta_n > S_{j-1} : X(\zeta_n) \in B_2, X(\zeta_{n-1}) \in B_1\},$$

$j \geq 1$. Then the jth passage time for the marked job is $P_j = T_j - S_{j-1}$. For passage times that are complete circuits in the network, $A_1 = B_1$ and $A_2 = B_2$ so that $S_j = T_j$.

It is intuitively clear and is shown in Appendix 2 that the sequence of passage times for any other job (as well as the sequence of passage times, irrespective of job identity, in order of start or termination) converges in distribution to the same random variable as the sequence of passage times for the marked job. The goal of the simulation is the estimation of $r(f) = E\{f(P)\}$, where f is a real-valued (measurable) function and P is the limiting passage time for the marked job.

Example 2.3. (Cyclic Queues With Feedback) Let $Z(t)$ be the number of jobs waiting or in service at center 1 at time t. The state space, G^*, of the augmented job stack process $X = \{X(t): t \geq 0\}$ is

$$G^* = \{(i,j): 0 \leq i \leq N, 1 \leq j \leq N\}.$$

Consider the passage time, P, that starts when a job completes service at center 2 (and joins the tail of the queue at center 1) and terminates when the job next joins the tail of the queue at center 2. The sets A_1 and A_2 defining the starts of this passage time for the marked job are

$$A_1 = \{(i,N): 0 \leq i < N\}$$

and

$$A_2 = \{(i,1): 0 < i \leq N\}.$$

The sets B_1 and B_2 defining the terminations of the passage time for the marked job are

$$B_1 = \{(i,i): 0 < i \leq N\}$$

and

$$B_2 = \{(i-1,i): 0 < i \leq N\}.$$

Now consider the passage time, R, that starts when a job completes service at center 2 (and joins the tail of the queue at center 1) and terminates the next such time the job joins the tail of the queue at center 1. This passage time is specified by the sets

$$A_1 = B_1 = \{(i,N): 0 \leq i < N\}$$

and

$$A_2 = B_2 = \{(i,1): 0 < i \leq N\}.$$

Figure 3.5 shows state transitions of the augmented job stack process and the four subsets A_1, A_2, B_1, and B_2 for $N = 2$ jobs. In the schematic representation of the job stack, jobs to the left of the

80 3 Markovian Networks

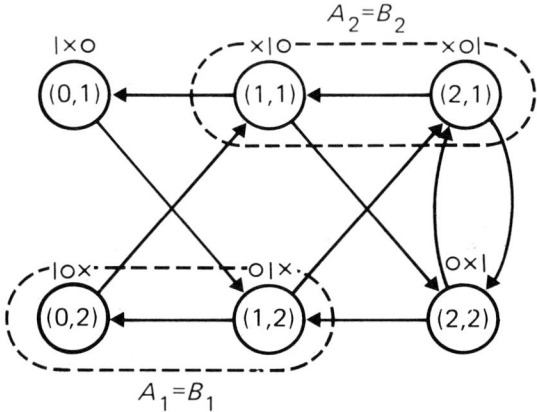

Figure 3.5. Subsets of G for passage time R

vertical bar are at center 1 and jobs to the right of the bar are at center 2. The symbol × denotes the marked job and ○ denotes an unmarked job.

Example 2.4. (Data Base Management System Model) Consider the passage time, R, that starts when a job completes service as class 7 at center 1 (and changes to class 6) and terminates the next time the job completes service at center 1 as class 7. This passage time is specified by the sets

$$A_1 = B_1 = \{(N - (i + 1), 0, 0, 0, 0, 0, 7, i, N - i): 0 \le i < N\}$$

and

$$A_2 = B_2 = \{(N - (i + 1), 0, 0, 0, 0, 0, 6, i, N - i): 0 \le i < N\}.$$

3.3 Irreducible, Closed Sets of Recurrent States

Proposition 1.9 asserts that the job stack process has a single irreducible, closed set of recurrent states provided that (i) the routing matrix P is irreducible and (ii) there is a service center that sees only one job class or is such that jobs of the lowest priority job class at the center are subject to preemption when a job of higher priority joins the queue. These conditions ensure that the job stack process has a limiting distribution independent of the initial state. Example 3.1 shows that the augmented job stack process can have more than one irreducible, closed set of recurrent states.

Example 3.1. Consider a network with one service center and two job classes and suppose that the set, C, of (center, class) pairs is $C = \{(1,1),(1,2)\}$. Suppose that jobs of class 1 have priority at center 1 over jobs of class 3 and that jobs of class 2 at center 1 are served in LCFS (last come, first served) order. Also suppose that the routing matrix P is

$$P = \begin{matrix} 0 & 1 \\ 1 & 0 \end{matrix}.$$

Set

(3.2) $$Z(t) = \left(C_2^{(1)}(t), C_1^{(1)}(t), S_1(t)\right),$$

where the class of the job in service at center 1 at time t is $S_1(t)$ and $C_j^{(1)}(t)$ is the number of jobs of class j waiting in queue at center 1, $j = 1,2$. With $N = 2$ jobs in the network, the state space, D^*, of the embedded jump chain $\{Z_k : k \geq 0\}$ is

$$D^* = \{(0,1,1),(0,1,2),(1,0,1),(1,0,2)\}.$$

Observe that $(0,1,2) \rightarrow (0,1,1)$, $(0,1,1) \rightarrow (1,0,1)$, $(1,0,1) \rightarrow (1,0,2)$, and $(1,0,2) \rightarrow (1,0,1)$, and that all other one-step transition probabilities are zero. States $(0,1,2)$ and $(0,1,1)$ are transient and the set, D, of recurrent states of $\{Z_k : k \geq 0\}$ is

$$D = \{(1,0,1),(1,0,2)\}.$$

The set, G, of recurrent states of the embedded jump chain $\{X(\zeta_k):k \geq 0\}$ is

$$G = \{(1,0,1,1),(1,0,2,1),(1,0,1,2),(1,0,2,2)\}.$$

It is easy to check that $\{(1,0,1,1),(1,0,2,1)\}$ and $\{(1,0,1,2),(1,0,2,2)\}$ of are irreducible subsets of G.

Proposition 3.4 asserts that for any network with more than one service center, the conditions of Proposition 1.9 ensure that the embedded jump chain $\{X(\zeta_k):k \geq 0\}$ has a single irreducible, closed set of recurrent states. For $x,x' \in G^*$, we write $x \to x'$ when the probability of transition from x to x' in one step of the embedded jump chain $\{(X(\zeta_n):n \geq 0\}$ is positive; we write $x \sim x'$ when x' is accessible from x: for some $n \geq 1$ the probability starting from x of entering x' on the nth step is positive.

The idea of the proof is to show that $x^* = (z^*,N)$ is accessible from any other state x of the augmented job stack process, where z^* is the state of the job stack process in which all N jobs are of class $j_{k(1)}(1)$ at center 1. (Without loss of generality we suppose that either $k(1) = 1$ or service to jobs of class $j_{k(1)}(1)$ at center 1 is subject to preemption.) By Proposition 1.9, state z^* is accessible from any $z \in D^*$; this implies that if $(z,n) \in G^*$, then $(z,n) \sim (z^*,n')$ for some n'. Therefore we need only show that $(z^*,n') \sim (z^*,N)$ for $1 \leq n' \leq N - 1$. For $0 < n < N$ and $(i,j) \in C$, let $z_n(i,j)$ be the state of the job stack process in which there are n jobs of class j at center i and $N - n$ jobs of class $j_{k(1)}(1)$ at center 1.

Lemma 3.3. Let $i_1 > 1$ and $(i_1,j_1) \in C$. Suppose that (i_2,j_2), $(i_3,j_3),...,(i_K,j_K) \in C$ are distinct (center, class) pairs such that $(i_K,j_K) = (1,j_{k(1)}(1))$ and $p_{i_m j_m, i_{m+1} j_{m+1}} > 0$ for $1 \leq m \leq K - 1$. Also suppose that $k(1) = 1$ or service at center 1 to a job of class $j_{k(1)}(1)$ is preempted when any other job of higher priority joins the queue. Then $(z_n(i_1,j_1), N - n) \sim (z_1^*,N)$ for all n ($1 \leq n \leq N$).

3.3 Irreducible, Closed Sets of Recurrent States 83

Proof. First suppose that $k(1) = 1$. Since $k(1) = 1$, $i_m > 1$ for $2 \leq m \leq K - 1$. Let l_1 $(1 < l_1 \leq K)$ be the index of the first (center, class) pair in $(i_2,j_2),...,(i_K,j_K)$ such that either the center is not i_1 or the center is i_1 and the class has lower priority (at center i_1) than class j_1. In state $(z_n(i_1,j_1), N - n)$, the marked job is in service at center 1 as class $j_{k(1)}(1)$ and a job of class j_1 is in service at center i_1. By the definition of l_1, either $i_{l_1} \neq i_1$ or $i_{l_1} = i_1$ and jobs of class j_{l_1} have lower priority at center i_1 than jobs of class j_1. Moreover, if $l_1 > 2$, then $i_l = i_1$ and jobs of class j_l have higher priority at center i_1 than jobs of class j_1 for $1 < l < l_1$. With positive probability the service time of the marked job exceeds the sum of the service times at center i_1 of the n jobs of class j_1 so that each of the jobs of class j_1 moves to center i_{l_1} and becomes class j_{l_1} before the marked job completes service at center i_1. Thus, $(z_n(i_1,j_1), N - n) \sim (z_n(i_{l_1},j_{l_1}), N - n)$ if $l_1 < K$ and $(z_n(i_1,j_1), N - n) \sim (z_1^*,N)$ if $l_1 = K$. Next let l_2 $(l_1 < l_2 \leq K)$ be the index of the first (center, class) pair in $(i_{l_1},j_{l_1}),...,(i_K,j_K)$ such that the center is not i_{l_1} or the center is i_{l_1} and the class has lower priority (at center i_{l_1}) than class j_{l_1}. By the same argument, the n jobs move to center i_{l_2} (as class j_{l_2}) and the other $N - n$ jobs remain at center 1 with the marked job in service so that $(z_n(i_{l_1},j_{l_1}), N - n) \sim (z_n(i_{l_2},j_{l_2}), N - n)$ if $l_2 < K$ and $(z_n(i_{l_1},j_{l_1}), N - n) \sim (z_1^*,N)$ if $l_2 = K$. Continuing in this way, the n jobs join the queue at center 1 before the marked job completes service, and $(z_n(i_1,j_1), N - n) \sim (z_1^*,N)$.

Now suppose that $k(1) > 1$ and service at center 1 to a job of class $j_{k(1)}(1)$ is preempted when any other job of higher priority joins the queue. As before, let l_1 $(1 < l_1 \leq K)$ be the index of the first (center, class) pair in $(i_2,j_2),...,(i_K,j_K)$ such that either the center is not i_1 or the center is i_1 and the class has lower priority (at center i_1) than class j_1. When the job stack process is in state $(z_n(i_1,j_1), N - n)$, the marked job is in service at center 1 as class $j_{k(1)}(1)$ and a job of class j_1 is in service at center i_1. Then we have $(z_n(i_1,j_1), N - n) \sim (z_n(i_{l_1},j_{l_1}), N - n)$ if $l_1 < K$ and $(z_n(i_1,j_1), N - n) \sim (z_1^*,N)$ if $l_1 = K$. (If $i_{l_1} > 1$ or if $i_{l_1} = 1$ and

$j_{l_1} = j_{k(1)}(1)$, this follows by the argument used for $k(1) = 1$. Otherwise, $i_{l_1} = 1$ and jobs of class j_{l_1} have higher priority at center 1 than jobs of class $j_{k(1)}(1)$. In this case $l_1 < K$ and with positive probability the (class j_1) job in service at center i_1 moves to center 1 as class j_{l_1} before the marked job completes service at center 1 and before another job (if any) completes service at center i_1. When this job moves to center 1, service at center 1 to the marked job is preempted and class j_{l_1} service for this job starts. Prior to completion of this class j_{l_1} service, with positive probability the remaining jobs at center i_1 move one at a time to center 1 as class j_{l_1}.) As before, let l_2 ($l_1 < l_2 \le K$) be the index of the first (center, class) pair in $(i_{l_1}, j_{l_1}), \ldots, (i_K, j_K)$ such that the center is not i_{l_1} or the center is i_{l_1} and the class has lower priority (at center i_{l_1}) than class j_{l_1}. By the same argument, $(z_n(i_{l_1}, j_{l_1}), N - n) \sim (z_n(i_{l_2}, j_{l_2}), N - n)$ if $l_2 < K$ and $(z_n(i_{l_1}, j_{l_1}), N - n) \sim (z_1^*, N)$ if $l_2 = K$. Continuing in this way, the n jobs join the queue at center 1 as class $j_{k(1)}$. Since a job that has been preempted receives additional service at the center before any other job of its class at the center receives service, $(z_n(i_1, j_1), N - n) \sim (z_1^*, N)$. □

Proposition 3.4. Suppose that the number of service centers $s > 1$. Also suppose that the routing matrix P is irreducible and for some service center, i_0, either $k(i_0) = 1$ or service to a job of class $j_{k(i_0)}(i_0)$ at center i_0 is preempted when any other job of higher priority joins the queue. Then the set, G, of recurrent states of the embedded jump chain $\{X(\zeta_k) : k \ge 0\}$ is irreducible.

Proof. Without loss of generality, suppose that $i_0 = 1$ and let $x \in G^*$. It is sufficient to show the existence of a state, x^*, of the augmented job stack process such that $x \sim x^*$ for all $x \in G^*$. Let z^* be the state of the job stack process in which there is one job of class $j_{k(1)}(1)$ in service at center 1 and $N - 1$ jobs of class $j_{k(1)}(1)$ in queue at center 1 (or in service if center 1 is a multiple server center.) Set $x^* = (z^*, N) \in G^*$. For $z \in D^*$ and $1 \le n \le N$, the argument in the proof of Proposition 1.9 shows that $z \sim z^*$ and therefore $(z,n) \sim (z^*, n')$ for some n'. If $n' = N$, we have $x = (z,n) \sim x^*$. Otherwise $1 \leq n' < N$, and we first show that $(z^*, n') \sim (z_k(i_{K+1}, j_{K+1}), N - k)$ for some $(i_{K+1}, j_{K+1}) \in C$ with $i_{K+1} > 1$ and some k ($1 \leq k < N$).

Since $s > 1$ and the routing matrix P is irreducible, there exists a finite sequence of distinct (center, class) pairs $(1, j_1)$, $(1, j_2), \ldots, (1, j_K)$, $(i_{K+1}, j_{K+1}) \in C$ such that $i_{K+1} > 1$ and $j_1 = j_{k(1)}(1)$ with $p_{1 j_K, i_{K+1} j_{K+1}} > 0$ and $p_{1 j_m, 1 j_{m+1}} > 0$ for $1 \leq m < K$. Since jobs of class $j_{k(1)}(1)$ have lowest priority at center 1, with positive probability in K steps the job in service at center 1 successively receives (center 1) service as class $j_1, j_2, \ldots, j_K$, and moves to center i_{K+1} as class j_{K+1} prior to completion of service to another job at center 1. Thus, $(z^*, n') \sim (z_1(i_{K+1}, j_{K+1}), n_1)$ for some $n_1 \leq N - 1$. In state $(z_1(i_{K+1}, j_{K+1}), n_1)$ the marked job is in service at center 1 if $n_1 = N - 1$ and is in queue at center 1 otherwise. If $n_1 < N - 1$, $(z_1(i_{K+1}, j_{K+1}), n_1) \sim (z_2(i_{K+1}, j_{K+1}), n_2)$ for some $n_2 \leq N - 2$ by the same argument. Continuing in this way for $k \leq N - 1$ steps, it follows that $(z^*, n') \sim (z_k(i_{K+1}, j_{K+1}), N - k)$.

We now show that $(z_k(i_{K+1}, j_{K+1}), N - k) \sim (z^*, N)$. Since the routing matrix P is irreducible, there exists a finite sequence of distinct (center, class) pairs (i_{K+2}, j_{K+2}), $(i_{K+3}, j_{K+3}), \ldots$, $(i_{K+L}, j_{K+L}) \in C$ such that $(i_{K+L}, j_{K+L}) = (1, j_{k(1)}(1))$, and $p_{i_m j_m, i_{m+1} j_{m+1}} > 0$ for $K + 1 \leq m < K + L$. First suppose that $L = 2$. With positive probability the job of class j_{K+1} in service at center i_{K+1} completes service and joins the queue at center 1 as class $j_{k(1)}(1)$ before the marked job completes service at center 1 so that $(z_k(i_{K+1}, j_{K+1}), N - k) \to (z_{k-1}(i_{K+1}, j_{K+1}), N - k + 1)$. If there are more jobs at center i_{K+1}, with positive probability each of these jobs goes into service at center i_{K+1}, completes service, and moves to center 1 while the marked job remains in service. Now suppose that $L > 2$. By assumption, either center 1 sees jobs of only one class ($k(1) = 1$) or service to a job of class $j_{k(1)}(1)$ is subject to preemption when jobs of higher priority join the queue at center 1. By Lemma 3.3, $(z_k(i_{K+1}, j_{K+1}), N - k) \sim (z^*, N)$. □

Corollary 3.5 is immediate.

Corollary 3.5. Suppose that the number of service centers $s > 1$. Also suppose that the routing matrix P is irreducible and for some service center, i_0, either $k(i_0) = 1$ or service to a job of class $j_{k(i_0)}(i_0)$ at center i_0 is preempted when any other job of higher priority joins the queue. Then restricted to the set, G, of recurrent states, the augmented job stack process $X = \{X(t): t \geq 0\}$ is irreducible and positive recurrent.

Example 3.6. (Data Base Management System Model) As $k(2) = 1$ and service to jobs of class $j_{k(1)}(1) = 7$ at center 1 is subject to preemption, either state $x_1^* = (N - 1,0,0,0,0,0,7,0,N)$ or $x_2^* = (0,0,0,0,0,0,0,N,N)$ can serve as a target state for the augmented job stack process. The set of recurrent states of $\{X(\zeta_k): k \geq 0\}$ is irreducible.

3.4 The Marked Job Method

Under the conditions of Proposition 3.4, the set, G, of recurrent states of the augmented job stack process X is irreducible. We assume that these conditions hold and that the subsets A_1, A_2, B_1, and B_2 that define passage times for the marked job are subsets of G. Point estimates and asymptotic confidence intervals for $r(f) = E\{f(P)\}$ can be obtained from a single simulation of the augmented job stack process by tracking a marked job.

Let X_n denote the state of the augmented job stack process when the $(n + 1)$st passage time for the marked job starts:

$$X_n = X(S_n),$$

$n \geq 0$. Since X is a CTMC and $\{S_n : n \geq 0\}$ are stopping times for the chain, $\{X_n : n \geq 0\}$ is a DTMC with finite state space, A_2. (We assume that the process $\{X_n : n \geq 0\}$ is aperiodic.) Furthermore, the process $\{(X_n, S_n) : n \geq 0\}$ satisfies

$$P\{X_{n+1} = j, S_{n+1} - S_n \leq t \mid X_0,\ldots,X_n; S_0,\ldots,S_n\}$$

$$= P\{X_{n+1} = j, S_{n+1} - S_n \leq t \mid X_n\}$$

with probability one for all $n \geq 0$, $j \in A_2$, and $t \geq 0$.

Proposition 4.1. The stochastic process $\{(X_n, S_n) : n \geq 0\}$ is a Markov renewal process.

This follows directly from the definition of a Markov renewal process (MRP). The basic data for this MRP is the semi-Markov kernel, $\{K(i,j;t) : i,j \in A_2, t \geq 0\}$, where

$$K(i,j;t) = P\{X_{n+1} = j, S_{n+1} - S_n \leq t \mid X_n = i\}.$$

While the kernel is normally given in the analysis of a MRP, for the network of queues passage time problem the kernel is virtually impossible to calculate. Thus from this point of view, the only hope is to generate sample paths of $\{(X_n, S_n) : n \geq 0\}$ via simulation of the augmented job stack process.

Select $x' \in A_2$ and begin the simulation of the augmented job stack process X with $X(0) = x'$. Carry out the simulation in cycles defined by the successive entrances of $\{X_n : n \geq 0\}$ to x'. Denote by α_k the length of (number of transitions in) the kth cycle of $\{X_n : n \geq 0\}$; the quantity α_k is the number of passage times for the marked job in the kth cycle. Set $\beta_0 = 0$ and $\beta_k = \alpha_1 + \ldots + \alpha_k$, $k \geq 1$. Also set

$$Y_1(f) = \sum_{j=1}^{\alpha_1} f(P_j)$$

and denote by $Y_m(f)$ the analogous summation over the mth cycle.

Proposition 4.2 is a consequence of Proposition 4.1 and the definition of a regenerative process. The random indices β_n such that $X(S_{\beta_n}) = x'$ form a sequence of regeneration points for the process $\{(X(S_n), P_{n+1}) : n \geq 0\}$; this follows from the fact that the sequence $\{P_n : n \geq 1\}$ starts from scratch when $X(S_{\beta_n}) = x'$. The expected time between regeneration points is finite since $\{X(S_n) : n \geq 0\}$ is an irreducible DTMC with finite state space.

Proposition 4.2. The process $\{(X(S_n), P_{n+1}) : n \geq 0\}$ is a regenerative process in discrete time and the expected time between regeneration points is finite.

Since $\{X(S_n) : n \geq 0\}$ is an aperiodic DTMC, it follows from Proposition 4.2 that $(X(S_n), P_{n+1}) \Rightarrow (X, P)$ as $n \to \infty$ and the regenerative property ensures that the pairs of random variables $\{(Y_k(f), \alpha_k) : k \geq 1\}$ are i.i.d.

Denote the set of discontinuities of the function f by $D(f)$. The proof of the ratio formula in Proposition 4.3 requires that $P\{P \in D(f)\} = 0$ but does not use the key renewal theorem.

Proposition 4.3. Provided that $P\{P \in D(f)\} = 0$ and $E\{|f(P)|\} < \infty$,

$$E\{f(P)\} = \frac{E\{Y_1(f)\}}{E\{\alpha_1\}}.$$

Proof. Assume that $f \geq 0$ and $E\{f(P)\} < \infty$. Set $f_c = \min(f, c)$ for c such that $0 < c < \infty$. Observe that $f_c(P_{n+1}) \Rightarrow f_c(P)$ as $n \to \infty$ and

(4.4) $$\lim_{n \to \infty} E\{f_c(P_{n+1})\} = E\{f_c(P)\}$$

since $P\{P \in D(f)\} = 0$ and f_c is bounded. Next compute the Césaro average of the sequence appearing in (4.4). We write

(4.5) $$\frac{E\left\{\sum_{n=0}^{m} f_c(P_{n+1})\right\}}{m+1} = \frac{E\left\{\sum_{k=1}^{l(m)+1} Y_k(f_c)\right\}}{m+1} - \frac{E\{Y'(m)\}}{m+1},$$

where $l(m) = \max\{k : \beta_k \le m\}$ and

$$Y'(m) = \sum_{n=m+1}^{\beta_{l(m)+1}} f_c(P_{n+1}).$$

Since $0 \le f_c \le c$, we have $0 \le Y'(m) \le c(\beta_{l(m)+1} - m)$. In addition, Wald's equation implies that

$$E\{\beta_{l(m)+1}\} = E\{\alpha_1\} E\{l(m) + 1\}$$

and

$$E\left\{\sum_{k=1}^{l(m)+1} Y_k(f_c)\right\} = E\{Y_1(f_c)\} E\{l(m) + 1\}.$$

These equations plus the elementary renewal theorem imply that

$$\lim_{n \to \infty} \frac{E\{\beta_{l(m)+1} - m\}}{m+1} = 0$$

and

$$\lim_{m \to \infty} \frac{E\left\{\sum_{k=1}^{l(m)+1} Y_k(f_c)\right\}}{m+1} = \frac{E\{Y_1(f_c)\}}{E\{\alpha_1\}}.$$

Hence from (4.5),

(4.6) $$\lim_{m \to \infty} \frac{E\left\{\sum_{n=0}^{m} f_c(P_{n+1})\right\}}{m+1} = \frac{E\{Y_1(f_c)\}}{E\{\alpha_1\}}.$$

From (4.4) and (4.6) we conclude that

(4.7) $$E\{f_c(P)\} = \frac{E\{Y_1(f_c)\}}{E\{\alpha_1\}}.$$

Now let $c \to \infty$ on both sides of (4.7) and use the assumption that

$E\{f(P)\} < \infty$ to obtain

(4.8) $$E\{f(P)\} = \frac{E\{Y_1(f)\}}{E\{\alpha_1\}}.$$

For a general f function, write $f = f^+ - f^-$ and apply the above argument to both f^+ and f^-. Thus (4.8) holds provided that $E\{|f(P)|\} < \infty$. □

With these results the standard regenerative method applies. Based on n cycles,

$$\hat{r}(n) = \frac{\bar{Y}(n)}{\bar{\alpha}(n)} = \frac{\sum_{m=1}^{n} Y_m(f)}{\sum_{m=1}^{n} \alpha_m(f)}$$

is a strongly consistent point estimate for $r(f)$ and an asymptotic $100(1 - 2\gamma)\%$ confidence interval is

$$\left[\hat{r}(n) - \frac{z_{1-\gamma} s(n)}{\bar{\alpha}(n) n^{1/2}}, \hat{r}(n) + \frac{z_{1-\gamma} s(n)}{\bar{\alpha}(n) n^{1/2}}\right].$$

The quantity $s^2(n)$ is a strongly consistent point estimate for $\sigma^2(f) = \text{var}(Y_1(f) - r(f)\alpha_1)$. Asymptotic confidence intervals for $r(f)$ are based on the c.l.t.

$$\frac{n^{1/2}\{\hat{r}(n) - r(f)\}}{\sigma(f)/E\{\alpha_1\}} \Rightarrow N(0,1)$$

as $n \to \infty$.

Algorithm 4.9. (Marked Job Method for Markovian Networks)

1. Select $x' \in A_2$ so that a passage time for the marked job starts when the augmented job stack process makes a transition from some state in A_1 to state x'.

2. Set $X(0) = x'$ and simulate the augmented job stack process. Observe a fixed number, n, of cycles defined by the successive times at which a passage time for the marked job starts and the augmented job stack process makes a transition to state x'. In each cycle measure the passage times for the marked job.
3. Compute the number, α_m, of passage times for the marked job in the mth cycle and the quantity

$$Y_m(f) = \sum_{j=\beta_{m-1}+1}^{\beta_m} f(P_j),$$

where $\beta_0 = 0$ and $\beta_m = \alpha_1 + \ldots + \alpha_m$.

4. Form the point estimate

$$\hat{r}(n) = \frac{\bar{Y}(n)}{\bar{\alpha}(n)}.$$

5. Form the asymptotic $100(1 - 2\gamma)\%$ confidence interval

$$\left[\hat{r}(n) - \frac{z_{1-\gamma} \, s(n)}{\bar{\alpha}(n) \, n^{1/2}}, \, \hat{r}(n) + \frac{z_{1-\gamma} \, s(n)}{\bar{\alpha}(n) \, n^{1/2}} \right].$$

Application of the marked job method requires the selection of a recurrent state $x' \in A_2$. Proposition 4.11 asserts that the recurrent states of the augmented job stack process X are of the form (z,n), where z is a recurrent state of the job stack process Z and $1 \leq n \leq N$. Let z_i^* be the state of the job stack process in which all N jobs are of class $j_{k(i)}(i)$ at center i, $1 \leq i \leq s$. Then, under the conditions of Proposition 3.4, $z \in D$ if and only if $z_i^* \sim z$ for some i ($i = 1,2,\ldots,s$).

Lemma 4.10. Suppose that the number of service centers $s > 1$. Also suppose that the routing matrix P is irreducible and for some service center, i_0, either $k(i_0) = 1$ or service to a job of class

$j_{k(i_0)}(i_0)$ at center i_0 is preempted when any other job of higher priority joins the queue. Then the set, $(z_{i_0}^*, n)$ is a recurrent state of the embedded jump chain $\{X(\zeta_k): k \geq 0\}$ for all n $(1 \leq n \leq N)$.

Proof. Without loss of generality, suppose that $i_0 = 1$: either $k(1) = 1$ or service to jobs of class $j_{k(1)}(1)$ at center 1 is subject to preemption. By Proposition 1.9, Observe that because (z_1^*, N) is accessible from all $(z,n) \in G^*$, $(z_1^*, N) \in G$ and it is sufficient to show that $(z^*, N) \sim (z_1^*, n)$ for all n, $1 \leq n \leq N - 1$. Since $s > 1$ and P is irreducible, there exists a finite sequence of (center, class) pairs $(i_1, j_1), (i_2, j_2), \ldots, (i_M, j_M) \in C$ such that $(i_1 j_1) = (i_M, j_M) = (1, j_{k(1)}(1))$, $p_{i_k j_k, i_{k+1} j_{k+1}} > 0$ for $1 \leq k \leq M - 1$, and $i_k > 1$ for some k with $2 \leq k \leq M - 1$. Let m be the largest index such that $i_m > 1$. Then there exists an index $k \leq m$ such that exactly one of the following holds: (i) $i_k > 1$ and $i_k \neq i_m$, (ii) $i_k = i_m$ and jobs of class j_m have equal or higher priority at center i_k than jobs of class j_k, or (iii) $i_{k-1} = 1$, $i_k = i_m$, and jobs of class j_k have higher nonpremptive priority at center i_k than jobs of class j_m. Let l be the smallest such index. Now fix n $(1 \leq n \leq N - 1)$. Starting in state (z_1^*, N), with positive probability the marked job completes service at center 1 (as class $j_{k(1)}(1)$) and moves to center i_m as class j_m while the other $N - 1$ jobs remain at center 1. Then, also with positive probability, $n - 1$ jobs complete service at center 1 and move to center i_l as class j_l while the marked job remains in service. (Recall that a job that has been preempted receives additional service at the center before any other job of its class at the center receives service.) Finally, with positive probability the marked job returns to center 1 (as class $j_{k(1)}(1)$) followed by the $n - 1$ jobs from center i_l before any other job completes service at center 1. Thus, $(z_1^*, N) \sim (z_1^*, n)$. □

Proposition 4.11. Under the conditions of Lemma 4.10, the set, G, of recurrent states of the embedded jump chain $\{X(\zeta_k): k \geq 0\}$ is $G = \{(z,n) \in G^* : z \in D\}$.

Proof. First suppose that $(z,m) \in G$. Then with probability one $\{X(\zeta_k): k \geq 0\}$ returns infinitely often to (z,m). This implies that with probability one $\{Z_k: k \geq 0\}$ returns infinitely often to z and that $z \in D$. Therefore, $G \subseteq \{(z,n): z \in D, 1 \leq n \leq N\}$.

Now let $z \in D$ and $(z,m) \in G^*$. By Lemma 4.10 there exists $z^* \in D$ such that $(z^*, n) \in G$ for all n, $1 \leq n \leq N$. Label the jobs from 1 to N and start the job stack process in state z^*. Let m_j be the position of job j in the job stack associated with state z^*, $1 \leq j \leq N$. By Proposition 1.4, with probability one the job stack process moves in a finite number of steps to state z. Let l_n be the label of the job in position n of the resulting job stack. Then $(z^*, m_{l_n}) \sim (z,m)$. Since $(z^*, m_{l_n}) \in G$ and only recurrent states are accessible from a recurrent state of a discrete time Markov chain, $(z,m) \in G$. □

Tables 3.1-3.3 give point and interval estimates for passage times in the cyclic queues with feedback model. Exponential service times have been generated by logarithmic transformation of uniform random numbers, $U_n = W_n/(2^{31} - 1)$, obtained from the recurrence relation

$$(4.12) \quad W_n = 7^5 W_{n-1} \pmod{2^{31} - 1} = 16807 \, W_{n-1} \pmod{2^{31} - 1}.$$

Independent streams of exponential random numbers (resulting from different seeds of the uniform random number generator) have been used to generate individual exponential holding time sequences. Table 3.1 contains estimates for characteristics of the passage time R in the cyclic queues with feedback model. The initial state for the augmented job stack process X (and return state identifying cycles) is the state $(1,1)$. The results in Table 3.1 are for $N = 2$ jobs, with $\mu_1 = 1$, $\mu_2 = 0.5$ and $p = 0.75$. (Service times at center i have mean μ_i^{-1}.) The theoretical value for the expected passage time is given in parentheses. In all cases, the 90% confidence intervals contain the theoretical value.

Table 3.1 Simulation Results for Passage Time R in Cyclic Queues With Feedback.

$N=2$, $\mu_1=1$, $\mu_2=0.5$, $p=0.75$.

Return State is (1,1).

	\multicolumn{6}{c}{No. of Cycles}				
	100	200	400	800	1000	2000
Simulated time	2606.44	5323.05	11647.92	23010.20	28541.88	57213.44
Fraction of time center 1 busy	0.8498	0.8483	0.8551	0.8498	0.8478	0.8544
Fraction of time center 2 busy	0.4237	0.4280	0.4171	0.4333	0.4384	0.4272
No. of transitions/ cycle (CTMC)	22.56	22.81	25.11	24.59	24.29	24.59
$E\{R\}$ (10.333)	10.050 ±0.6608	10.274 ±0.4496	10.447 ±0.3332	10.271 ±0.2250	10.282 ±0.1993	10.402 ±0.1476
$P\{R \le 2.333\}$	0.0799 ±0.0254	0.0714 ±0.0174	0.0714 ±0.0123	0.0725 ±0.0087	0.0719 ±0.0077	0.0721 ±0.0054
$P\{R \le 4.667\}$	0.2813 ±0.0436	0.2596 ±0.0296	0.2620 ±0.0213	0.2579 ±0.0149	0.2540 ±0.0131	0.2572 ±0.0090
$P\{R \le 10.333\}$	0.6076 ±0.0481	0.5993 ±0.0324	0.6067 ±0.0225	0.6201 ±0.0161	0.6195 ±0.0148	0.6174 ±0.0103
$P\{R \le 14.0\}$	0.8021 ±0.0407	0.8136 ±0.0276	0.8013 ±0.0143	0.8110 ±0.0131	0.8120 ±0.0117	0.8084 ±0.0083
$P\{R \le 18.667\}$	0.9271 ±0.0240	0.9129 ±0.0196	0.8970 ±0.0143	0.9081 0.0095	0.9073 ±0.0086	0.9012 ±0.0064

The estimates for the passage time P in Tables 3.2 and 3.3 illustrate the effect of different return states; e.g., 2367 transitions in the CTMC were required for 100 cycles defined by entrances to the state (1,1), but only 1183 transitions for entrances to the state (2,1). Since entrances to state (2,1) occur approximately twice as frequently as entrances to state (1,1), we would expect that only half as many cycles would be needed for comparable accuracy; this is borne out by the results. In all cases the 90% confidence intervals contain the theoretical value.

Table 3.2 Simulation Results for Passage Time P in Cyclic Queues With Feedback.

$N=2$, $\mu_1=1$, $\mu_2=0.5$, $p=0.75$.

Return State is $(1,1)$.

	\multicolumn{5}{c}{No. of Cycles}			
	100	200	400	800	1000
Simulated time	2747.95	5463.94	11751.03	21413.70	27355.86
Fraction of time center 1 busy	0.8501	0.8465	0.8557	0.8507	0.8485
Fraction of time center 2 busy	0.4203	0.4305	0.4170	0.4312	0.4366
No. of transitions/ cycle (CTMC)	23.67	23.31	25.33	22.97	23.27
$E\{P\}$ (6.667)	6.448 ±0.5668	6.594 ±0.3830	6.820 ±0.2895	6.570 ±0.2178	6.584 ±0.1950
$P\{P \leq 1.667\}$	0.2119 ±0.0360	0.2105 ±0.0268	0.2068 ±0.0192	0.2180 ±0.0134	0.2122 ±0.0123
$P\{P \leq 3.333\}$	0.4073 ±0.0410	0.3887 ±0.0306	0.3773 ±0.0212	0.3878 ±0.0161	0.3826 ±0.0142
$P\{P \leq 6.667\}$	0.6457 ±0.0419	0.6180 ±0.0308	0.6138 ±0.0203	0.6290 ±0.0157	0.6317 ±0.0143
$P\{P \leq 10.000\}$	0.8013 ±0.0346	0.7843 ±0.0244	0.7699 ±0.0179	0.7833 ±0.0137	0.7830 ±0.0125
$P\{P \leq 13.333\}$	0.8940 ±0.0282	0.8879 ±0.0196	0.8656 ±0.0156	0.8733 ±0.0126	0.8736 ±0.0103

Table 3.4 gives estimates for the passage time R in the data base management system model. Class 1 service time is an exponentially distributed random variable, B, and class j service time is an exponentially distributed random variable, A_j, $j = 2,3,...,7$. The initial state for the augmented job stack process is $(N - 2,0,0,0,0,0,6,1,N - 1)$.

Table 3.3 Simulation Results for Passage Time P
in Cyclic Queues With Feedback.

$N=2$, $\mu_1=1$, $\mu_2=0.5$, $p=0.75$.

Return State is (2,1).

	\multicolumn{5}{c}{No. of Cycles}				
	100	200	400	800	1000
Simulated time	1333.29	2727.80	5730.14	11074.78	137810.08
Fraction of time center 1 busy	0.8511	0.8537	0.8483	0.8563	0.8563
Fraction of time center 2 busy	0.4134	0.4161	0.4272	0.4163	0.4193
No. of transitions/ cycle (CTMC)	11.83	11.75	10.21	11.94	11.87
$E\{P\}$ (6.667)	6.414 ±0.6865	6.656 ±0.6144	6.723 ±0.4474	6.754 ±0.3233	6.739 ±0.2852
$P\{P \leq 1.667\}$	0.1974 ±0.0537	0.1919 ±0.0370	0.1987 ±0.0278	0.2044 ±0.0203	0.2029 ±0.0180
$P\{P \leq 3.333\}$	0.3618 ±0.0615	0.3434 ±0.0447	0.3711 ±0.0331	0.3774 ±0.0239	0.3778 ±0.0210
$P\{P \leq 6.667\}$	0.5855 ±0.0621	0.5993 ±0.0463	0.6125 ±0.0324	0.6234 ±0.0232	0.6290 ±0.0206
$P\{P \leq 10.000\}$	0.7829 ±0.0544	0.7980 ±0.0411	0.7833 ±0.0288	0.7863 ±0.0205	0.7835 ±0.0182
$P\{P \leq 13.333\}$	0.8816 ±0.0399	0.8822 ±0.0305	0.8686 ±0.0422	0.8634 ±0.0170	0.8625 ±0.0182

Comparison of Table 3.4 with Table 3.1 reveals the effect on simulation running time of the considerable structural complexity of the data base management system model. For the return states chosen, there are approximately the same number of transitions in the CTMC for 250 cycles of this model as for 400 cycles of the cyclic queues with feedback model. The 90% confidence intervals for the expected passage time contain the theoretical values.

Table 3.4 Simulation Results for Passage Time R in Data Base Management System Model.

$N=2$, $E\{B\}=50.0$, $E\{A_6\}=6.7$, $E\{A_2\}=E\{A_4\}=3.3$,

$E\{A_3\}=E\{A_5\}=1.5$, $E\{A_7\}=1.0$, $p_1=0.1$, $p_2=0.2$.

Return state is (0,0,0,0,0,0,6,1,1).

	No. of Cycles					
	100	200	400	800	1000	2000
Simulated time	3707.05	6061.72	14054.86	20771.41	35803.24	426610.02
Fraction of time center 1 busy	0.7485	0.8351	0.7640	0.7441	0.7404	0.7577
Fraction of time center 2 busy	0.6247	0.5435	0.6072	0.6147	0.6132	0.5966
No. of transitions/ cycle (CTMC)	42.76	38.46	40.95	43.00	41.04	41.69
$E\{R\}$ (84.556)	102.974 ±24.638	87.851 ±16.533	95.611 ±11.237	810.136 ±11.837	87.969 ±10.125	83.501 ±8.657
$P\{R \leq 75\}$	0.5555 ±0.1152	0.5942 ±0.0855	0.5918 ±0.0817	0.6198 ±0.0474	0.6241 ±0.0416	0.6399 ±0.0359
$P\{R \leq 100\}$	0.5555 ±0.1152	0.6522 ±0.0871	0.6667 ±0.0790	0.6916 ±0.0487	0.6929 ±0.0420	0.7182 ±0.0367
$P\{R \leq 125\}$	0.7222 ±0.1139	0.7681 ±0.0746	0.7483 ±0.0736	0.7485 ±0.0466	0.7543 ±0.0399	0.7730 ±0.0343
$P\{R \leq 150\}$	0.7778 ±0.1094	0.8261 ±0.0606	0.8163 ±0.0623	0.8024 ±0.0440	0.8034 ±0.0382	0.8024 ±0.0391
$P\{R \leq 175\}$	0.8333 ±0.0859	0.8551 ±0.0557	0.8435 ±0.0562	0.8413 ±0.0400	0.8403 ±0.0355	0.8413 ±0.0309

3.5 Fully Augmented Job Stack Processes

We now consider the steady state estimation problem for passage times that correspond to passage through a subnetwork of a specified network of queues. The labelled jobs method prescribes simulation of a "fully augmented job stack process" that maintains the position of each of the jobs in the linear job stack. With this estimation procedure observed passage times for all the jobs enter into the point and interval estimates.

The labelled jobs method is based on simulation of the fully augmented job stack process restricted to an arbitrarily selected irreducible, closed set of recurrent states. (The set of recurrent states of the fully augmented job stack process need not be irreducible even if the set of recurrent states of the job stack process is irreducible.) We consider CTMC's obtained by restricting the fully augmented job stack process to distinct irreducible, closed sets of recurrent states. When the job stack process restricted to its recurrent states is irreducible, these CTMC's have a common generator, Q. A consequence is that the corresponding sequences of passage times (irrespective of job identity and enumerated in start order) converge to a common random variable.

Label the jobs from 1 to N and let $N^i(t)$ be the position of job i in the job stack at time t, $i = 1,2,...,N$. Then in terms of the vector $Z(t)$ of (1.1), set

$$(5.1) \qquad X^0(t) = (Z(t), N^1(t), ..., N^N(t)).$$

The process $X^0 = \{X^0(t): t \geq 0\}$ is called the *fully augmented job stack process*. The discussion of the labelled jobs method is for networks of queues having single server service centers. To handle networks having multiple server service centers, it is sufficient to into the job stack the class of job being served by each of the servers at a multiple server center.

The labelled jobs method prescribes simulation of the fully augmented job stack process in random blocks defined by the terminations of distinguished passage times that (i) terminate when no other passage times are underway and (ii) leave a fixed configuration of the job stack. These terminations serve to decompose the sequence of passage times for all of the jobs into i.i.d. blocks. In order for condition (i) to be satisfied, an additional

3.5 Fully Augmented Job Stack Processes

assumption is needed. Set

(5.2) $\quad S = \{(x_1, x_2) : x_1 \in A_1, x_2 \in A_2 \text{ and } x_1 \rightarrow x_2\}$

and

(5.3) $\quad T = \{(x_1, x_2) : x_1 \in B_1, x_2 \in B_2 \text{ and } x_1 \rightarrow x_2\}.$

(Recall that we write $x \rightarrow x'$ when there is a positive probability that the embedded jump chain $\{X(\zeta_n) : n \geq 0\}$ makes a transition in one step from state x to state x'.) We assume that $S \cap T = \emptyset$.

Proposition 5.4 is a consequence of the assumptions of Markovian routing and exponential service times.

Proposition 5.4. The fully augmented job stack process $X^0 = \{X^0(t) : t \geq 0\}$ is a continuous time Markov chain with finite state space, G^0.

Example 5.5 shows that the set of recurrent states of the fully augmented job stack process X^0 need not be irreducible even if the job stack process has a single irreducible, closed set of recurrent states.

Example 5.5. Suppose that the set C of (center, class) pairs is $C = \{(1,1),(2,1)\}$ and that the routing matrix P is

$$P = \begin{matrix} 0 & 1 \\ 1 & 0 \end{matrix}.$$

Also supppose that all queue service disciplines are FCFS. Set $Z(t) = Q_1(t)$, where $Q_1(t)$ is the number of jobs waiting or in service at center 1 at time t. When there are $N = 3$ jobs in the network, all states of the job stack process Z and the augmented job stack process X are recurrent, but the fully augmented job stack process X^0 has two irreducible, closed sets of recurrent states:

$$\{(z,1,2,3),(z,2,3,1),(z,3,1,2) : z = 0,1,2,3\}$$

and

$$\{(z,1,3,2),(z,3,2,1),(z,2,1,3):z = 0,1,2,3\}.$$

The recurrent states of any DTMC can be divided in a unique manner into irreducible, closed sets. As in Example 5.5, even if the set of recurrent states of the job stack process Z is irreducible, the fully augmented job stack process X^0 can have more than one irreducible, closed set of recurrent states. For each such set there is a sequence of passage times (irrespective of job identity and enumerated in order of passage time start). In Section 3.6 we show that when the set, D, of recurrent states of the job stack process Z is irreducible, these sequences converge in distribution to a common random variable, P^0, the limiting passage time for any marked job.

A consequence of this result is that if f is a real-valued (measurable) function with domain $[0, +\infty)$, point estimates and confidence intervals for $r^0(f) = E\{f(P^0)\}$, can be obtained by restricting a single simulation of the fully augmented job stack process X^0 to any irreducible, closed set of recurrent states. We also show in Section 3.7 that if the set D is irreducible, then $(z,n^1,...,n^N)$ is a recurrent state of the fully augmented job stack process X^0 for all $z \in D$ and distinct $n^1,...,n^N$ with $1 \le n^1,...,n^N \le N$.

3.6 The Labelled Jobs Method

From now on assume that the set D is irreducible. Also suppose that the subsets A_1, A_2, B_1, and B_2) that define the starts and terminations of passage times are subsets of the set $G = \{(z,n): z \in D, 1 \le n \le N\}$. Select a recurrent state z^0 of the job stack process such that (i) a passage time for some job terminates when the process hits state z^0 and (ii) no other passage times are underway. (We assume that such a z^0 exists.) Also select $(z^0,n^1,...,n^N) \in G^0$ and set $X^0(0) = (z^0,n^1,...,n^N)$; thus we consider

3.6 The Labelled Jobs Method

simulation of the process X^0 restricted to some irreducible, closed set of recurrent states, G_1^0. Denote by $\{P_n^0:n \geq 1\}$ the successive passage times (irrespective of job identity) enumerated in termination order. Also, let $\{T_n^0:n \geq 1\}$ be the corresponding sequence of termination times. Observe the fully augmented job stack process X^0 for a fixed number of random blocks defined by the successive entrances of $\{X^0(T_n^0):n \geq 1\}$ to the set of states $\{(z,m^1,...,m^N) \in G_1^0:z = z^0\}$.

Let α_m^0 be the length (in discrete time units) of the mth block of $\{X(T_n^0):n \geq 1\}$, $m \geq 1$. (The quantity α_m^0 is the number of passage times irrespective of job identity in the mth block.) Next, let $Y_m^0(f)$ be the sum of the quantities $f(P_j^0)$ over the passage times P_j^0 in the mth block of the fully augmented job stack process X^0; e.g.,

$$Y_1^0(f) = \sum_{j=1}^{\alpha_1^0} f(P_j^0).$$

Propositions 6.1 and 6.2 lead to point and interval estimates for $r^0(f)$. Recall that $D(f)$ denotes the set of discontinuities of the function f.

Proposition 6.1. The sequence of pairs of random variables $\{(Y_m^0(f),\alpha_m^0):m \geq 1\}$ are independent and identically distributed.

Standard arguments establish a ratio formula for $r^0(f)$.

Proposition 6.2. Provided that $P\{P \in D(f)\} = 0$ and $E\{|f(P^0)|\} < \infty$,

$$r^0(f) = \frac{E\{Y_1^0(f)\}}{E\{\alpha_1^0\}}.$$

With these results

$$\hat{r}^0(n) = \frac{\bar{Y}^0(n)}{\bar{\alpha}^0(n)} = \frac{\sum_{m=1}^{n} Y_m^0(f)}{\sum_{m=1}^{n} \alpha_m^0}$$

is a strongly consistent point estimate for $r^0(f)$ and an asymptotic $100(1 - 2\gamma)\%$ confidence interval is

$$\left[\hat{r}^0(n) - \frac{z_{1-\gamma} s^0(n)}{\bar{\alpha}^0(n) n^{1/2}}, \hat{r}^0(n) + \frac{z_{1-\gamma} s^0(n)}{\bar{\alpha}^0(n) n^{1/2}} \right].$$

The quantity $(s^0(n))^2$ is a strongly consistent point estimate for $(\sigma^0(f))^2 = \text{var}(Y_1^0(f) - r^0(f)\alpha_1^0)$. Asymptotic confidence intervals for $r^0(f)$ are based on the c.l.t.

$$\frac{n^{1/2} \{\hat{r}^0(n) - r^0(f)\}}{\sigma^0(f)/E\{\alpha_1^0\}} \Rightarrow N(0,1)$$

as $n \to \infty$.

Algorithm 6.3. (Labelled Jobs Method for Markovian Networks)

1. Select a recurrent state, z^0, of the job stack process such that a passage time for some job terminates and no other passage times are underway when the process makes a transition to state z^0. Also select an initial state $(z^0, n^1, ..., n^N)$ for the fully augmented job stack process.

2. Set $X^0(0) = (z^0, n^1, ..., n^N)$ and simulate the fully augmented job stack process. Observe a fixed number, n, of blocks defined by the successive times $\{T_{\beta_k}^0 : k \geq 1\}$ at which a passage time terminates and the job stack process makes a transition to state z^0. In each block measure the passage times for all of the jobs.

3. Compute the number, α_m^0, of passage times in the mth block and the quantity

$$Y_m^0(f) = \sum_{j=\beta_{m-1}+1}^{\beta_m} f(P_j^0).$$

4. Form the point estimate

$$\hat{r}^0(n) = \frac{\bar{Y}^0(n)}{\bar{\alpha}^0(n)}.$$

5. Form the asymptotic $100(1 - 2\gamma)\%$ confidence interval

$$\left[\hat{r}^0(n) - \frac{z_{1-\gamma}\, s^0(n)}{\bar{\alpha}^0(n)\, n^{1/2}},\, \hat{r}^0(n) + \frac{z_{1-\gamma}\, s^0(n)}{\bar{\alpha}^0(n)\, n^{1/2}}\right].$$

3.7 Sequences of Passage Times

We now consider sequences of passage times obtained by restricting the fully augmented job stack process X^0 to distinct irreducible, closed sets of recurrent states and show that these sequences converge to a common random variable. As illustrated by Example 5.5, distinct irreducible, closed sets of recurrent states of the process X^0 differ only in the labelling of the jobs in each position of the stack; because all jobs (irrespective of label) are stochastically identical, the process X^0 restricted to any irreducible, closed set of recurrent states has the same stochastic structure.

Let $Z(t)$ be the vector of (1.1) associated with the job stack at time t. As in Section 3.5, label the jobs from 1 to N and set

(7.1) $$X^0(t) = (Z(t), N^1(t), \ldots, N^N(t)),$$

where $N^j(t)$ is the position of the job labelled j in the job stack at time t, $j = 1, 2, \ldots, N$. It is convenient to introduce an equivalent specification of the position in the network of each job. Set

(7.2) $$W^0(t) = (Z(t), N_1(t), ..., N_N(t)),$$

where $N_k(t)$ is the label of the job in position k of the job stack at time t, $k = 1, 2, ..., N$. Denote the state space of the processes $X^0 = \{X^0(t) : t \geq 0\}$ and $W^0 = \{W^0(t) : t \geq 0\}$ by G^0 and H^0, respectively. The equivalence of the two specifications is in terms of a one-to-one mapping, ψ, of G^0 onto H^0: for $x = (z, n^1, ..., n^N) \in G^0$

(7.3) $$\psi(z, n^1, ..., n^N) = (z, n_1, ..., n_N),$$

where $n_k = j$ when $n^j = k$ (i.e., job k is in position j of the job stack), $k = 1, 2, ..., N$. According to this definition, $\psi(x) \in H^0$ since $1 \leq n_1, ..., n_N \leq N$ and the n_k are distinct. (If $n_k = n_l = m$ then $n^m = k$ and $n^m = l$.) It is easy to see by the same argument that the mapping ψ is invertible, and thus is one-to-one and onto. Our strategy is to obtain the desired results for the process W^0 and then use the mappings ψ and ψ^{-1} to extend them to the process X^0.

Definition 7.4. Let ϕ be a permutation of the set $\{1, 2, ..., N\}$. For $w = (z, n_1, ..., n_N) \in H^0$, $(z, \phi(n_1), ..., \phi(n_N))$ is called the *image of w under ϕ* and is denoted by $\phi(w)$. For $x = (z, n^1, ..., n^N) \in G^0$, the *image of x under ϕ* is $(z, \phi(n^1), ..., \phi(n^N))$ and is denoted by $\phi(x)$.

For $w, w' \in H^0$, we write $w \to w'$ when the probability of transition in one step of the embedded jump chain of W^0 from w to w' is positive, and write $w \sim w'$ when w' is accessible from w: for some $n \geq 1$ the probability starting from w of entering w' on the nth step is positive.

Lemma 7.5. Denote by $R = \{r(w, w') : w, w' \in H^0\}$ the jump matrix of the continuous time Markov chain W^0 and let $q = \{q(w) : w \in H^0\}$ be the vector of rate parameters for the holding times in W^0. Let ϕ be a permutation of the set $\{1, 2, ..., N\}$. Then

$$q(w) = q(\phi(w))$$

and

$$r(w,w') = r(\phi(w),\phi(w'))$$

for all $w,w' \in H^0$. Moreover, $\phi(w) \sim \phi(w')$ whenever $w \sim w'$.

Proof. Let $w = (z,n_1,...,n_N) \in H^0$. First observe that $q(w)$ does not depend on the labels of the jobs in the various positions of the job stack associated with z. Since $\phi(w) = (z,\phi(n_1),...,\phi(n_N))$, $q(w) = q(\phi(w))$.

We now show that $r(w,w') = r(\phi(w),\phi(w'))$ for all $w = (z,n_1,...,z_N)$, $w' = (z',n'_1,...,n'_N) \in H^0$. Denote by σ_j the index of n_j in $(n'_1,...,n'_N)$: $n'_{\sigma_j} = n_j$ for $j = 1,2,...,N$; σ_j is the position in the job stack associated with z' of the job that was in position j of the job stack associated with state z. By definition of the induced mapping, $\phi(w) = (z,\phi(n_1),...,\phi(n_N))$, and $\phi(w') = (z,\phi(n'_1),...,\phi(n'_N))$. Let σ'_j be the index of $\phi(n_j)$ in $(\phi(n'_1),...,\phi(n'_N))$. Since $n'_{\sigma_j} = n_j$, $\phi(n'_{\sigma_j}) = \phi(n_j)$. Hence $\sigma'_j = \sigma_j$ for each j; i.e., the one-step transition from z to z' takes the job in any position of the job stack associated with z to the same position in the job stack associated with z', irrespective of how the jobs are labelled. The jump probability $r(\phi(w),\phi(w'))$ is equal to the conditional probability that the embedded jump chain of the job stack process Z makes a transition from z to z' in such a way that the job labelled $\phi(n_j)$ when the configuration of the job stack is z is in position σ'_j when the configuration of the job stack is z', $j = 1,2,...,N$; this is exactly the jump probability $r(w,w')$.

Finally we show that $\phi(w) \sim \phi(w')$ whenever $w \sim w'$. If $w \sim w'$, there exists a sequence $w_0,w_1,...,w_n \in J_1^0$ such that $w_0 = w$, $w_n = w'$, and $r(w_i,w_{i+1}) > 0$, $i = 0,1,...,n-1$. But by the result just proved, $r(\phi(w_i),\phi(w_{i+1})) > 0$ also, and therefore $\phi(w) \sim \phi(w')$. □

Proposition 7.6. Suppose that the set, D, of recurrent states of the job stack process $Z = \{Z(t): t \geq 0\}$ is irreducible. Let J_1^0, J_2^0 be irreducible, closed sets of recurrent states of the process W^0 and denote by $Q_i = \{q_i(w,w'): w,w' \in J_i^0\}$ the generator of the continuous time Markov chain restricted to the set J_i^0, $i = 1,2$. Then there exists a one-to-one mapping, ϕ_{12}, of J_1^0 onto J_2^0. Moreover, $q_1(w,w') = q_2(\phi_{12}(w),\phi_{12}(w'))$ for all $w,w' \in J_1^0$.

Proof. Select $z^* \in D$. Since D is irreducible, there exist $n_1^*(i)$, $n_2^*(i),...,n_N^*(i)$ such that $(z^*,n_1^*(i),...,n_N^*(i)) \in J_i^0$, $i = 1,2$. Let ϕ_{12} be the permutation of $\{1,2,...,N\}$ that takes $(n_1^*(1),...,n_N^*(1))$ into $(n_1^*(2),...,n_N^*(2))$. Then the induced mapping, ϕ_{12}, given by

$$\phi_{12}(z,n_1,...,n_N) = (z,\phi_{12}(n_1),...,\phi_{12}(n_N))$$

for all $w = (z,n_1,...,n_N) \in J_1^0$, is a one-to-one mapping of J_1^0 onto J_2^0. Observe that since J_1^0 is irreducible, $(z^*,n_1^*(1),...,n_N^*(1)) \sim w$ for all $w \in J_1^0$ and thus $\phi_{12}(w) \in J_2^0$. By the argument in Lemma 7.5, we have $(z^*,\phi_{12}(n_1^*(1)),...,\phi_{12}(n_N^*(1))) \sim \phi_{12}(w)$; i.e., $(z^*,n_1^*(2),...,n_N^*(2)) \sim \phi_{12}(w)$ so that $\phi_{12}(w) \in J_2^0$. The mapping is one-to-one since it is one-to-one on $(n_1,...,n_N)$. To see that the mapping is onto, let $w' = (z,n_1',...,n_N') \in J_2^0$. We claim that $w = \phi_{12}^{-1}(w') \in J_1^0$. (Observe that $(z^*,n_1^*(2),...,n_N^*(2)) \sim w'$. By the last part of Lemma 7.5, $\phi_{12}^{-1}(z^*,n_1^*(2),...,n_N^*(2)) \sim \phi_{12}^{-1}(w')$ and $\phi_{12}^{-1}(z^*,n_1^*(2),...,n_N^*(2)) = (z^*,n_1^*(1),...,n_N^*(1))$ by definition. It follows that $(z^*,n_1^*(1),...,n_N^*(1)) \sim \phi_{12}^{-1}(w')$. Since $\phi_{12}^{-1}(w') = w$, $w \in J_1^0$.)

Now set $R^1 = \{r(w,w'): w,w' \in J_1^0\}$ and $q^1 = \{q(w): w \in J_1^0\}$. Also set $R^2 = \{r(\phi_{12}(w),\phi_{12}(w')): w,w' \in J_1^0\}$ and $q^2 = \{\phi_{12}(w): w \in J_1^0\}$. Since the jump matrix and vector of rate parameters for holding times determine the infinitesimal generator of a CTMC, it is sufficient to show that $R^1 = R^2$ and $q^1 = q^2$. These results follow from Lemma 7.5. □

We can now show that the CTMC's obtained by restricting the fully augmented job stack process X^0 to an irreducible, closed set of recurrent states have a common generator.

Proposition 7.7. Suppose that the set, D, of recurrent states of the job stack process Z is irreducible. Let G_1^0 and G_2^0 be irreducible, closed sets of recurrent states of the fully augmented job stack process X^0 and $Q^i = \{q_i(\psi(x),\psi(x')):x,x' \in G_i^0\}$ be the generator of the continuous time Markov chain X^0 restricted to the set G_i^0, $i = 1,2$. Then there exists a one-to-one mapping ζ_{12} of G_1^0 onto G_2^0 such that

$$q_1(x,x') = q_2(\zeta_{12}(x),\zeta_{12}(x'))$$

for all $x,x' \in G_1^0$.

Proof. Consider $\zeta_{12} = \psi^{-1}\phi_{12}\psi$, where ψ_{12} is the mapping of Proposition 7.5. Let $J_i^0 = \{\psi(x):x \in G_i^0\}$, $i = 1,2$. Since ψ is a one-to-one onto mapping and the sets G_2^0 and J_2^0 are finite, $G_1^0 = \{\psi^{-1}(w):w \in G_2^0\}$. It follows (since ϕ_{12} is a one-to-one onto mapping) that ζ_{12} is a one-to-one mapping of G_1^0 onto G_2^0.

Let $\{q_i^*(w,w'):w,w' \in J_i^0\}$ be the generator of the continuous time Markov chain W^0 restricted to the set J_i^0, $i = 1,2$. By Proposition 7.6, $q_1^*(w,w') = q_2^*(\phi_{12}(w),\phi_{12}(w'))$. Hence, $q_1(x,x') = q_2(\psi^{-1}\phi_{12}\psi(x),\psi^{-1}\phi_{12}\psi(x')) = q_2(\zeta_{12}(x),\zeta_{12}(x'))$ for all $x,x' \in G_1^0$. □

Proposition 7.8 provides a characterization of the recurrent states of the fully augmented job stack process X^0. The proof uses three properties of a DTMC with finite state space: (i) at least one state is recurrent, (ii) any state accessible from a recurrent state is recurrent, and (iii) any transient state reaches a recurrent state.

Proposition 7.8. Suppose that the set, D, of recurrent states of the job stack process $Z = \{Z(t):t \geq 0\}$ is irreducible. Then the set of recurrent states of the fully augmented job stack process X^0 is $\{(z,n^1,n^2,...,n^N) \in G^0 : z \in D\}$.

Proof. Denote by J^0 the set of recurrent states of the process W^0. Since $x \in G^0$ if and only if $\psi(x) \in J^0$, it is sufficient to show that $J^0 = \{(z,n_1,n_2,...,n_N) \in H^0 : z \in D\}$. First observe that no element of the set $\{(z,n^1,...,n^N) \in G^0 : z \notin D\}$ is a recurrent state of the process X^0 for if X^0 returns infinitely often to $(z,n^1,...,n^N)$ with probability one, then Z returns infinitely often to z with probability one and consequently $z \in D$.

Now let $z \in D$ and $x = (z, n^1, \ldots, n^N) \in G^0$. Since G^0 is finite, X^0 has at least one recurrent state. Moreover, there exists $x^* \in G^0$ such that x^* is recurrent and $x \sim x^*$. Denote the configuration of the job stack associated with x^* by z^*. Since x^* is recurrent, $z^* \in D$, and $z^* \sim z$ because D is irreducible. Then there exist $m^1, \ldots, m^N$ such that $x^* \sim (z, m^1, \ldots, m^N)$. Let γ be the permutation of the set $\{1, 2, \ldots, N\}$ that takes $(n^1, \ldots, n^N)$ into $(m^1, \ldots, m^N)$. Necessarily there exists l such that γ^l is the identity permutation.

The argument in the proof of Lemma 7.5 shows that $\phi(x) \sim \phi(x')$ whenever $x \sim x'$, where ϕ is any permutation of $\{1, 2, \ldots, N\}$ and $\phi(x) = (z, \phi(n^1), \ldots, \phi(n^N))$ is the image of x under ϕ. In particular, $\gamma(z, n^1, \ldots, n^N) \sim \gamma(z, m^1, \ldots, m^N)$ if $(z, n^1, \ldots, n^N) \sim (z, m^1, \ldots, m^N)$. Iterating this, we have $(z, n^1, \ldots, n^N) \sim (z, m^1, \ldots, m^N) = (z, \gamma(n^1), \ldots, \gamma(n^N))$, $(z, \gamma(n^1), \ldots, \gamma(n^N)) \sim (z, \gamma^2(n^1), \ldots, \gamma^2(n^N)), \ldots$, $(z, \gamma^{l-1}(n^1), \ldots, \gamma^{l-1}(n^N)) \sim (z, \gamma^l(n^1), \ldots, \gamma^l(n^N))$. Since x^* is recurrent and $x^* \sim (z, \gamma(n^1), \ldots, \gamma(n^N))$, $(z, \gamma(n^1), \ldots, \gamma(n^N))$ is recurrent, and so are $(z, \gamma^2(n^1), \ldots, \gamma^2(n^N)), \ldots, (z, \gamma^l(n^1), \ldots, \gamma^l(n^N))$. It follows that $(z, n^1, \ldots, n^N)$ is recurrent since $(z, \gamma^l(n^1), \ldots, \gamma^l(n^N)) = (z, n^1, \ldots, n^N)$. □

We now show that the sequences of passage times (irrespective of job identity) associated with irreducible, closed sets of recurrent states of the fully augmented job stack process X^0 converge to a common random variable.

Proposition 7.9. Set $X(t) = (Z(t), N(t))$ and suppose that the set of recurrent states of $X = \{X(t) : t \geq 0\}$ is irreducible. Let $z_0 \in D$. Denote the initial state of the fully augmented job stack process X^0 by $(z_0, n^1, \ldots, n^N)$ and let $\{P_k^0 : k \geq 1\}$ be the successive passage times (irrespective of job identity) enumerated in start order. Then $P_k^0 \Rightarrow P^0$ as $k \to \infty$ for all $(z_0, n^1, \ldots, n^N) \in G^0$, where P^0 is the limiting passage time for any marked job.

Proof. Let $x_1 = (z_0, n_1^1, \ldots, n_1^N) \in G^0$. By Proposition 7.8 $x_1 \in G_1^0$, an irreducible, closed set of recurrent states for the process X^0. Set $X^0(0) = x_1$ and let $\{P_k^i(1): k \geq 1\}$ be the sequence of passage times enumerated in start order for job i, $i = 1, 2, \ldots, N$. We show below that the sequences $\{P_k^i(1): k \geq 1\}$ converge in distribution to a common random variable, P^0. Then without change, the second part of the argument used to establish Proposition 2.1 of Appendix 2 shows that $P_k^0(1) \Rightarrow P^0$ as $k \to \infty$.

Consider $x_2 = (z_0, n_2^1, \ldots, n_2^N)$. If $x_2 \in G_1^0$ there is nothing more to prove. Otherwise, let G_2^0 be the irreducible, closed set of recurrent states for the process X^0 that contains x_2. For $X^0(0) = x_2$, let $\{P_k^i(2): k \geq 1\}$ be the successive passage times in start order for job i. Let $j(i)$ be the job label such that $n_2^{j(i)} = n_1^i$. Then $P_k^{j(i)}(2)$ and $P_k^i(1)$ are identically distributed, $k \geq 1$. This is a consequence of Proposition 7.7 using the one-to-one mapping, ζ_{12}, of G_1^0 onto G_2^0 for which $\zeta_{12}(x_1) = x_2$. It follows that $P_k^i(2) \Rightarrow P^0$ as $k \to \infty$ for all i and also that $P_k^0(2) \Rightarrow P^0$ as $k \to \infty$.

We now show that $P_k^i(1) \Rightarrow P^0$ as $k \to \infty$ for all i. Set $X^0(0) = x_1$ and, without loss of generality, assume that job 1 is the first job for which a passage time starts after time 0 and that this passage time starts at time S_0^1. Denote $Z(S_0^1)$ by z^1 and $N^1(S_0^1)$ by n^1. Let $\{S_k^1: k \geq 0\}$ be the sequence of passage time starts for job 1. Then the process $\{(Z(S_k^1), N^1(S_k^1), P_{k+1}^1): k \geq 0\}$ is a regenerative process in discrete time. It follows that $\{P_{k+1}^1: k \geq 0\}$ converges in distribution to a limit, P^0. Now consider the analogous processes $\{(Z(S_k^i), N^i(S_k^i), P_{k+1}^i): k \geq 0\}$, $i = 2, 3, \ldots, N$. By assumption, the set of recurrent states of the process $\{(Z(t), N^i(t)): t \geq 0\}$ is irreducible; moreover, (z^1, n^1) is a recurrent state of the process. Therefore, with probability one there exists a time $S_{k_i}^i$ at which the job labelled i starts a passage time and the job is in position n^1 of the job stack associated with state z_1. Then, element by element, the two sequences $\{(Z(S_k^i), N^i(S_k^i), P_{k+1}^i): k \geq k_i\}$ and $\{(Z(S_k^1), N^1(S_k^1), P_{k+1}^1): k \geq 0\}$ are identically distributed. It follows that $P_k^i \Rightarrow P^0$ as $k \to \infty$. □

3.8 Networks with Multiple Job Types

Previously we have discussed estimation methods for networks of queues with priorities among job classes and stochastically identical jobs. We now consider networks with multiple job types. These are networks with stochastically nonidentical jobs; the type of a job may influence its routing through the network as well as its service requirements at each center. For expository convenience, we assume that there are only two job types in the network.

We consider closed networks of queues with a finite number, N, of jobs of two types, and assume that there are N_1 jobs of type 1 and N_2 jobs of type 2 with $N_1 + N_2 = N$. There are a finite number of service centers, s, and a finite number of job classes, c. As before, we denote the set of (center, class) pairs in the network by C. All jobs retain their job type, but may change class as they traverse the network. (Think of type 1 jobs as cubes and type 2 jobs as spheres, and let job classes correspond to different colors. We permit jobs to change color, but not shape.) Upon completion of service at center i, a type v job of class j goes to center k and changes to class l with probability $p_{ij,kl}^{(v)}$, where

$$P^{(v)} = \{p_{ij,kl}^{(v)} : (i,j), (k,l) \in C\}$$

is a given irreducible stochastic matrix.

The service times and service disciplines at each service center are as in Section 3.1 with the exception that they may also depend on job type. We briefly review the situation. At each service center jobs queue and receive service according to a fixed priority scheme among classes and types, where the scheme can vary from center to center. Each center operates as a single server, processing jobs of a fixed type and class according to a fixed service discipline. All service times in the network are mutually independent, and at each center have an exponential distribution with parameter that may depend on the service center, type and class of job being serviced, and the state of the entire system. A job in service may or may not be preempted (according to a fixed procedure for each center) if another job of higher priority joins the queue at the center.

To characterize the state of the system at time t, let $S_i(t)$ be the (type, class) pair of the job receiving service at center i at time t, $i = 1,2,...,s$; by convention $S_i(t) = (0,0)$ if there are no jobs at center i at time t. Denote by $j_1(i),...,j_{k(i)}(i)$ the (type, class) pairs served at center i ordered by decreasing priority, and let $C^{(i)}_{j_1}(t),...,C^{(i)}_{j_{k(i)}}(t)$ be the number of jobs in queue at time t of the various (type, class) pairs served at center i. We order the N jobs in a linear stack and set

(8.1) $Z(t) = \left(C^{(1)}_{j_{k(1)}}(t),...,C^{(1)}_{j_1}(t), S_1(t); ...; C^{(s)}_{j_{k(s)}}(t),...,C^{(s)}_{j_1}(t), S_s(t) \right).$

The job stack at time t again corresponds to the nonzero components in the vector $Z(t)$. Within a (type, class) pair at a center, jobs waiting appear in the job stack in the order of arrival at the center, the latest to arrive being closest to the top of the stack. Proposition 8.2 is a direct consequence of the assumptions of Markovian job routing and exponential service times.

Proposition 8.2. The job stack process $Z = \{Z(t) : t \geq 0\}$ is a continuous time Markov chain with finite state space, D^*.

Example 8.3 shows that irreducibility of the routing matrices $P^{(\nu)}$ does not ensure that the set of recurrent states of the embedded jump chain $\{Z_k : k \geq 0\}$ is irreducible.

Example 8.3. Consider a closed network of queues with one service center, two job classes, and two job types. Suppose that the set, C, of (center, class) pairs is $C = \{(1,1),(1,2)\}$. Also suppose that the routing matrices $P^{(\nu)}$ are

$$P^\nu = \begin{matrix} & 0 & 1 \\ & 1 & 0 \end{matrix}$$

and that type ν jobs of class 1 have nonpreemptive priority at center 1 over jobs of class 2, $\nu = 1,2$. Assume that the (type, class) pairs served at center 1 ordered by decreasing priority are $j_1(1) = (1,1)$, $j_2(1) = (1,2)$, $j_3(3) = (2,1)$, and $j_4(1) = (2,2)$, and that all queue service disciplines are FCFS. Set

$$Z(t) = \left(C^{(1)}_{(2,2)}(t), C^{(1)}_{(2,1)}(t), C^{(1)}_{(1,2)}(t), C^{(1)}_{(1,1)}(t), S_1(t)\right),$$

where $S_1(t)$ is the (type, class) pair in service at center 1 at time t and $C^{(1)}_{(\nu,j)}(t)$ is the number of type ν jobs of class j in queue at cetner 1 at time t, $j = 1,2$. With $N = 2$ jobs, the sets $\{(1,0,0,0,1,1),(1,0,0,0,1,2)\}$ and $\{(0,1,0,0,1,1),(0,1,0,0,1,2)\}$ of recurrent states of the embedded jump chain $\{Z_k : k \geq 0\}$ are irreducible.

Proposition 8.6 asserts that the set of recurrent states of the embedded jump chain $\{Z_k : k \geq 0\}$ is irreducible provided that (i) the routing matrices $P^{(\nu)}$ are irreducible and (ii) there is a service center that sees only one job class ($k(i_0) = 2$ for some center i_0). The idea of the proof is to show the existence of a target state of the job stack process Z that is accessible from any state $z \in D^*$.

Lemma 8.4. Suppose that the routing matrices $P^{(\nu)}$ are irreducible and that $k(1) = 2$. Also suppose that type 1 jobs have higher priority than type 2 jobs at center 1. Let z_ν^* be the state of the job stack process in which all N jobs are at center 1 and there is a type ν job in service. Then $z_2^* \sim z_1^*$.

Proof. Denote by $(\nu, j(\nu))$ the lowest priority (type, class) pair at center 1 for type ν jobs, $\nu = 1,2$. Because the routing matrix $P^{(2)}$ is irreducible, there exists a finite sequence of (center, class) pairs $(i_1,j_1),(i_2,j_2),\ldots,(i_M,j_M)$ such that $(i_1,j_1) = (i_M,j_M) = (1,j(2))$ and $p^{(2)}_{i_m j_m, i_{m+1} j_{m+1}} > 0$ for $m = 1,2,\ldots,M-1$. Since $k(i) = 2$, we may suppose that $i_m > 1$ for $m = 2,\ldots,M-1$. Denote by z_k the state in which there is a type 1 job in service at center 1, $N_1 - 1$ type 1 jobs and $N_2 - 1$ type 2 jobs in queue at center 1, and one type 2 job of class j_k in service at center i_k, $k = 1,2,...,M - 1$. Then $z_2^* \sim z_1^*$ since $z_2^* \to z_1$, $z_1 \to z_2,..., z_{M-2} \to z_{M-1}$, and $z_{M-1} \to z_1^*$. □

Definition 8.5. Let $z, z' \in D^*$. Then z' is a *neighbor* of z if and only if the job stacks associated with z and z' are the same except that for unique $(i,j),(k,l) \in C$ with $p_{ij,kl}^{(v)} > 0$ for some v, in the job stack associated with z' there is one more type v job of class l at center k and one less type v job of class j at center i.

Proposition 8.6. Suppose that the routing matrices $P^{(v)}$ are irreducible and that $k(i_0) = 2$ for some service center i_0. Then the set of recurrent states of the embedded jump chain $\{Z_k : k \geq 0\}$ is irreducible.

Proof. By Lemma 3.2 it is sufficient for irreducibility to show the existence of a state z_1^* such that $z \sim z_1^*$ for all $z \in D^*$. Without loss of generality, assume that $k(1) = 2$ and that $j_{k(1)}(1) = (2, j(2))$, where $(v, j(v))$ is the lowest priority (type, class) pair at center 1 for jobs of type v, $v = 1,2$. Since the routing matrices $P^{(v)}$ are irreducible, there exist finite sequences $(i_1^{(v)}, j_1^{(v)})$, $(i_2^{(v)}, j_2^{(v)}),..., (i_{M(v)}^{(v)}, j_{M(v)}^{(v)}) \in C$ such that (i) $(i_1^{(v)}, j_1^{(v)}) = (i_{M(v)}^{(v)}, j_{M(v)}^{(v)}) = (1, j(v))$, (ii) for any (center, class) pair $(i,j) \in C$, there exists $n(v)$ $(1 \leq n(v) \leq M(v))$ such that $(i_{n(v)}^{(v)}, j_{n(v)}^{(v)}) = (i,j)$, and (iii) $p_{i_m j_m, i_{m+1} j_{m+1}}^{(v)} > 0$ for $m = 1,2,...,M(v) - 1$. Let $l_m(v)$ be the index of the first occurrence of $(1, j(v))$ following (i_m, j_m) in $(i_1^{(v)}, j_1^{(v)}), (i_2^{(v)}, j_2^{(v)}),...,(i_{M(v)}^{(v)}, j_{M(v)}^{(v)})$:

$$l_m(v) = \min \{l \geq m : (i_l^{(v)}, j_l^{(v)}) = (1, j(v))\},$$

$m = 1,2,...,M(v) - 1$. Then for $(i,j) \in C$ set

$$m_v(i,j) = 1 + \min \{m : (i_m^{(v)}, j_m^{(v)}) = (i,j) \text{ and } l_m - m \leq l_n - n$$

$$\text{for all } n \text{ such that } (i_n^{(v)}, j_n^{(v)}) = (i,j)\}$$

and in terms of this index define the type ν successor (center, class) pair, $s_\nu(i,j)$, of (i,j) as $s_{\nu(i,j)} = (i_{m_{\nu(i,j)}}, j_{m_{\nu(i,j)}})$.

Let $U(z)$ be the set of all (center, class) pairs $(i,j) \in C - \{(1,j(1)),(1,j(2))\}$ such that there is a job (either type 1 or type 2) of class j in service at center i when the job stack process is in state z. Let h be a function taking values in C and having domain $D^* \times \{1,2,...,N\}$ such that for $z \in D^*$ and $n \in \{1,2,...,N\}$ the value of $h(z,n)$ is (i,j) when the job in position n of the job stack associated with state z is of class j at center i. Define a nonnegative distance from z to z_1^* as follows. For the type ν job in position n of the job stack associated with state z, set

$$d(z,n;z_1^*) = \min\{l_m(\nu) - m : (i_m^{(\nu)}, j_m^{(\nu)}) = h(z,n)\},$$

$n = 1,2,...,N$. Then the distance, $d(z;z_1^*)$, from z to z_1^* is

$$d(z,z_1^*) = \sum_{n=1}^{N} d(z,n;z_1^*).$$

According to this definition, $d(z,z_2^*) = d(z,z_1^*)$ for all $z \in D^*$.

First suppose that $z \neq z_1^*, z_2^*$ and observe that $U(z)$ is nonempty since $k(1) = 2$. Select $(k,l) \in U(z)$ and let z_1 be the neighbor of z having one more type $\nu(1)$ job of class $j_{m_{\nu(1)}(k,l)}$ at center $i_{m_{\nu(1)}(k,l)}$ and one less type $\nu(1)$ job of class l at center k, where $\nu(1)$ is the type of the job of class l in service at center k. It follows from the definition of the successor (center, class) pair $s_{\nu(1)}(k,l)$ of (k,l) that $z \rightarrow z_1$. Moreover, $d(z_1;z_1^*) < d(z;z_1^*)$. If $d(z_1,z_1^*) > 0$, then $U(z_1)$ is nonempty. Select $(k_1,l_1) \in U(z_1)$ and let z_2 be the neighbor of z_1 having one more type $\nu(2)$ job of (center, class) pair $s_{\nu(2)}(k_1,l_1)$ and one less job of class l_1 at center k_1. Clearly, $z_1 \rightarrow z_2$ and necessarily $d(z_2;z_1^*) < d(z_1;z_1^*)$. Continuing in this way for at most a finite number, n, of steps, the distance to z_1^* decreases to zero and either $z_n = z_1^*$ or $z_n = z_2^*$. It follows using (Lemma 8.4 if $z_n = z_2^*$) that $z \sim z_1^*$. Next suppose that $z = z_2^*$ and conclude that $z \sim z_1^*$ by Lemma 8.4. Finally, suppose that $z = z_1^*$ so that there exists $z' \neq z_1^*$ such that $z^* \rightarrow z'$. By the previous argument, $z \sim z_1^*$. □

An argument similar to that used in the proof of Proposition 8.6 shows that the set of recurrent states of the embedded jump chain is irreducible if there exists a service center, i_0, such that the lowest priority (type, class) pair at center i_0 for type 1 and type 2 jobs are $j_{k(i_0)-1}(i_0)$ and $j_{k(i_0)}(i_0)$ and service to each of these (type, class) pairs at center i_0 is preempted when any other job of higher priority joins the queue. Corollary 8.7 is immediate since the job stack process has a finite state space.

Corollary 8.7. Suppose that the routing matrices $P^{(\nu)}$ are irreducible and that $k(i_0) = 2$ for some service center i_0. Then restricted to the set, D, of recurrent states, the job stack process $Z = \{Z(t): t \geq 0\}$ is irreducible and positive recurrent.

Proposition 8.8 is a direct consequence of Corollary 8.7 and the definition of a regenerative process.

Proposition 8.8. Under the conditions of Proposition 8.6, the job stack process $Z = \{Z(t): t \geq 0\}$ is a regenerative process in continuous time and the expected time between regeneration points is finite.

Since the state space of the job stack process is discrete and the expected time between regeneration points is finite, $Z(t) \Rightarrow Z$ as $t \to \infty$ by Proposition 2.3 of Chapter 2. A strongly consistent point estimate and asymptotic confidence intervals for $r(f) = E\{f(Z)\}$ can be obtained as in Section 3.1.

3.9 Simulation for Passage Times

We now consider estimation of passage times in networks with multiple job types and mark one job of each type. By tracking these two jobs, strongly consistent point estimates and asymptotic confidence intervals for a variety of passage time characteristics can be produced. The estimation procedure can also be applied to networks with only a single job type; the result is an alternative estimation scheme to that proposed in Section 3.6. When the passage time is a complete circuit or loop in a closed network, we refer to it as a response time.

As in Section 3.5, we view the N jobs as being completely ordered in a linear stack and set

$$Z(t) = (C^{(1)}_{j_{k(1)}}(t),\ldots,C^{(1)}_{j_1}(t),S_1(t);\ldots;C^{(s)}_{j_{k(s)}}(t),\ldots,C^{(s)}_{j_1}(t),S_s(t)).$$

The linear stack again corresponds to the order of components in the vector $Z(t)$ after ignoring any zero components. Within a (type, class) pair at a center, jobs waiting appear in the linear stack in the order of their arrival in the center, the latest to arrive being closest to the top of the stack. Let $N_\nu(t)$ be the position (from the top) of the type ν marked job in the job stack at time t. Then set

$$X(t) = (Z(t), N_1(t), N_2(t)).$$

Under the exponential service time and Markovian routing assumptions, the augmented job stack process $X = \{X(t): t \geq 0\}$ is a CTMC with finite state space, $G.^*$.

We assume that the process X has a single irreducible, closed set of recurrent states, G, and specify the passage (or response) times for the two types of jobs by eight nonempty subsets of G: $A_1^{(\nu)}$, $A_2^{(\nu)}$, $B_1^{(\nu)}$, $B_2^{(\nu)}$, $\nu = 1,2$. The sets $A_1^{(\nu)}$ and $A_2^{(\nu)}$ [resp., $B_1^{(\nu)}$, $B_2^{(\nu)}$] determine when to start [resp., stop] the clock measuring a particular passage time for the type ν marked job. Denoting the jump times of X by $\{\zeta_n: n \geq 0\}$, for $k, n \geq 1$ we require that the sets $A_1^{(\nu)}$, $A_2^{(\nu)}$, $B_1^{(\nu)}$ and $B_2^{(\nu)}$ satisfy:

if $X(\zeta_{n-1}) \in A_1^{(\nu)}$, $X(\zeta_n) \in A_2^{(\nu)}$, $X(\zeta_{n-1+k}) \in A_1^{(\nu)}$ and $X(\zeta_{n+k}^{(\nu)}) \in A_2^{(\nu)}$, then $X(\zeta_{n-1+m}) \in B_1^{(\nu)}$ and $X(\zeta_{n+m}) \in B_2^{(\nu)}$ for some $0 < m \leq k$;

and

if $X(\zeta_{n-1}) \in B_1^{(\nu)}$, $X(\zeta_n) \in B_2^{(\nu)}$, $X(\zeta_{n-1+k}) \in B_1^{(\nu)}$ and $X(\zeta_{n+k}^{(\nu)}) \in B_2^{(\nu)}$, then $X(\zeta_{n-1+m}) \in A_1^{(\nu)}$ and $X(\zeta_{n+m}) \in A_2^{(\nu)}$ for some $0 \leq m < k$.

3.9 Simulation for Passage Times

These conditions ensure that the start and termination times for the specified passage time strictly alternate. (We assume that for all $x \in A_2^{(\nu)}$ there exists $x \in A_1^{(\nu)}$ such that $x_1 \to x_2$ and that for all $x_2 \in B_2^{(\nu)}$ there exists $x_1 \in B_1^{(\nu)}$ such that $x_1 \to x_2$.) Also in terms of the jump times of the augmented job stack process, we define four sequences of random times: $\{S_j^{(\nu)} : j \geq 0\}$ and $\{T_j^{(\nu)} : j \geq 1\}$, $\nu = 1, 2$. The start [resp., termination] time of the jth passage time for the type ν marked job is denoted by $S_{j-1}^{(\nu)}$ [resp., $T_j^{(\nu)}$]. Formally, we have

$$S_j^{(\nu)} = \inf\{\zeta_n \geq T_j^{(\nu)} : X(\zeta_n) \in A_2^{(\nu)}, X(\zeta_{n-1}) \in A_1^{(\nu)}\}$$

and

$$T_j^{(\nu)} = \inf\{\zeta_n > S_{j-1}^{(\nu)} : X(\zeta_n) \in B_2^{(\nu)}, X(\zeta_{n-1}) \in B_1^{(\nu)}\},$$

$j \geq 1$. The jth passage time for the type ν marked job is $P_j^{(\nu)} = T_j^{(\nu)} - S_{j-1}^{(\nu)}$. For response times of type ν jobs, $A_1^{(\nu)} = B_1^{(\nu)}$, $A_2^{(\nu)} = B_2^{(\nu)}$, and $S_j^{(\nu)} = T_j^{(\nu)}$ for all $j \geq 1$.

Let $L(t)$ be the last state visited by the process X before jumping to $X(t)$ and set

$$V(t) = (L(t), X(t)).$$

The process $V = \{V(t) : t \geq 0\}$ has a state space, F, consisting of all pairs of states (x, x') for which a transition in X from state x to state x' can occur with positive probability. In general, of course, the size of the state space F is larger than that of G. The generator of the CTMC V can be obtained easily from that of X. Since X has a single, irreducible, closed set of recurrent states, so does V. Clearly, the entrance times of V to a state $(x, x') \in F$ correspond to the times of transition in X from state x to state x'. For a type ν job, we define two subsets of F according to:

$$S^{(\nu)} = \{(x_1, x_2) \in F: x_1 \in A_1^{(\nu)}, x_2 \in A_2^{(\nu)}\}$$

and

$$T^{(\nu)} = \{(x_1, x_2) \in F: x_1 \in A_1^{(\nu)}, x_2 \in B_2^{(\nu)}\}.$$

Thus, the entrances of V to $S^{(\nu)}$ [resp., $T^{(\nu)}$] correspond to the start [resp., termination] times of passage times for the type ν marked job. For response times of a type ν job, $S^{(\nu)} = T^{(\nu)}$.

The argument employed in Appendix 2 shows that $P_n^{(\nu)} \Rightarrow P^{(\nu)}$ as $n \to \infty$ and that the sequence of passage times for any other job of type ν also converges in distribution to the same random variable $P^{(\nu)}$. Moreover, the sequence of passage times of type ν jobs (irrespective of job identity) in the order of start (or termination) also converges in distribution to $P^{(\nu)}$. Our concern is with the estimation of characteristics of these limiting passage times.

Estimation of $E\{R^{(1)}\}$ and $P\{R^{(1)} \le x\}$

We first consider the estimation of characteristics of a limiting response time, $R^{(1)}$, of a type 1 job. For this estimation problem, of course, it is not necessary to mark a type 2 job. Since $R^{(1)}$ is a response time, $S^{(1)} = T^{(1)}$. We select a fixed element of $S^{(1)}$ that for convenience we designate state 0, and assume that $V(0) = 0$.

Suppose first that we wish to estimate $E\{R^{(1)}\}$. The successive entrances of the process V to $S^{(1)}$ constitute the starts and terminations of response times of the type 1 marked job. Let $R_n^{(1)}$ be the time between the nth and $(n+1)$st entrances to $S^{(1)}$, with the 0th entrance to $S^{(1)}$ occurring at $t = 0$. Also, let $\{V_n : n \ge 0\}$ denote the embedded jump chain associated with V. The random times $\{\tau_n : n \ge 1\}$ and $\{\delta_n : n \ge 1\}$ denote the lengths of the successive 0-cycles (successive returns to the fixed state 0) for V and $\{V_n : n \ge 0\}$, respectively. Then the number of response times for the type 1 marked job in the first 0-cycle of V is

$$N_1^{(1)} = \sum_{n=0}^{\delta_1-1} 1_{\{V_n \in S^{(1)}\}}$$

and the sum of the response times in that cycle is simply

$$\tau_1 = \sum_{n=0}^{N_1^{(1)}-1} R_n^{(1)}.$$

We denote the analogous quantities in the kth 0-cycle by $N_k^{(1)}$ and τ_k. Since V is a regenerative process, the pairs of random variables $\{(\tau_k, N_k^{(1)}): k \geq 1\}$ are i.i.d., and a renewal argument shows that

(9.1) $$E\{R^{(1)}\} = \frac{E\{\tau_1\}}{E\{N_1^{(1)}\}}$$

provided that $E\{R^{(1)}\} < \infty$. At this point, the arguments leading to the standard regenerative method apply. Set $\sigma^2 = \text{var}(\tau_1 - E\{R^{(1)}\}N_1^{(1)})$. Based on n cycles we can construct the strongly consistent point estimate $\bar{\tau}/\overline{N}^{(1)}$ and an associated asymptotic confidence interval for $E\{R^{(1)}\}$ if an estimate is available for σ. Asymptotic confidence intervals are obtained from the c.l.t.

$$\frac{n^{1/2}\left[\bar{\tau}/\overline{N}^{(1)} - E\{R^{(1)}\}\right]}{\sigma/E\{N_1^{(1)}\}} \Rightarrow N(0,1)$$

as $t \to \infty$.

To estimate $P\{R^{(1)} \leq x\}$, proceed as above, but also define the i.i.d. sequence of random variables $\{Y_k: k \geq 1\}$, where, e.g.,

$$Y_1 = \sum_{n=1}^{N_1^{(1)}} 1_{\{R_n^{(1)} \leq x\}}.$$

Then a strongly consistent point estimate for $P\{R^{(1)} \le x\}$ is $\overline{Y}/\overline{N}^{(1)}$, and we obtain asymptotic confidence intervals in the usual way.

Estimation of $E\{R^{(1)}\}$ and $E\{R^2\}$

Suppose that we wish to estimate the expected response time for type 2 jobs, $E\{R^{(2)}\}$, as well as $E\{R^{(1)}\}$. Response times for the type 2 marked job start and terminate at the entrance times of V to the set $S^{(2)} = T^{(2)}$. Let $N_k^{(2)}$ be the number of entrances to $S^{(2)}$ of V in the kth 0-cycle. For example, in the first 0-cycle

$$N_1^{(2)} = \sum_{n=0}^{\delta_1 - 1} 1_{\{V_n \in S^{(2)}\}}.$$

Although we are able to begin the simulation at the start of a response time for the type 1 marked job, in general a response time for the type 2 marked job is underway at $t = 0$. Similarly, at the end of a 0-cycle, a response time for the type 1 marked job terminates, but a response time for the type 2 marked job is still underway. After n 0-cycles, $N_1^{(2)} + \ldots + N_n^{(2)}$ response times for the type 2 marked job have started and the sum of these response times is approximately $\tau_1 + \ldots + \tau_n$. The error in this approximation is due to the partial response time at $t = 0$ that is not counted in $N_1^{(2)} + \ldots + N_n^{(2)}$ and the last response time that is counted, but does not terminate before the end of the nth 0-cycle. Since the point estimates and confidence intervals here are based on large sample theory (strong laws and central limit theorems), these errors are negligible for n large. In fact, the errors due to the two response times at $t = 0$ and at the end of the simulation run compensate for each other. Again, we have i.i.d. pairs of random variables $\{(\tau_k, N_k^{(2)}): k \ge 1\}$ and the ratio formula

(9.2) $$E\{R^{(2)}\} = \frac{E\{\tau_1\}}{E\{N_1^{(2)}\}},$$

provided that $E\{R^{(2)}\} < \infty$. A strongly consistent point estimate for $E\{R^{(2)}\}$ is $\bar{\tau}/\bar{N}^{(2)}$, and we can use the standard regenerative method to obtain an asymptotic confidence interval.

Estimation of $E\{R^{(1)}\} - E\{R^{(2)}\}$

Suppose now that we wish to estimate $r^{(1)} - r^{(2)}$, where $r^{(1)} = E\{R^{(1)}\}$ and $r^{(2)} = E\{R^{(2)}\}$. We can take as a point estimate the quantity $(\bar{\tau}/\bar{N}^{(1)}) - (\bar{\tau}/\bar{N}^{(2)})$, but need a bivariate c.l.t. in order to produce a confidence interval. To this end, set

$$Z_k^{(\nu)} = \tau_k - r^{(\nu)} N_k^{(\nu)}$$

and

$$Z_k = (Z_k^{(1)}, Z_k^{(2)}),$$

$k \geq 1$. (We take all our vectors to be column vectors.) The random vectors $\{Z_k : k \geq 1\}$ are i.i.d. since each Z_k is only a function of the kth 0-cycle. Furthermore, (9.1) and (9.2) imply that $E\{Z_k\} = 0$. Denoting the transpose of Z_k by Z_k', let $\Sigma = E\{Z_k Z_k'\} = \{\sigma_{ij}\}$ be the covariance matrix of the Z_k's. Assuming that the elements of Σ are finite, we have the c.l.t.

(9.3) $$n^{-1/2} \sum_{k=1}^{n} Z_k \Rightarrow N(0, \Sigma)$$

as $n \to \infty$, where $N(0, \Sigma)$ is a multivariate normal random variable with zero mean vector and covariance matrix Σ. We can rewrite (9.3) in the form

(9.4) $$n^{1/2} \begin{bmatrix} \bar{N}^{(1)}\{(\bar{\tau}/\bar{N}^{(1)}) - r^{(1)}\} \\ \bar{N}^{(2)}\{(\bar{\tau}/\bar{N}^{(2)}) - r^{(2)}\} \end{bmatrix} \Rightarrow N(0, \Sigma)$$

as $n \to \infty$. Since $\bar{N}^{(\nu)} \to E\{N_1^{(\nu)}\}$ with probability one, we can replace $\bar{N}^{(1)}$ and $\bar{N}^{(2)}$ outside the braces by $E\{N_1^{(1)}\}$ and $E\{N_1^{(2)}\}$ in (9.4) and not change the result. This is an application of the continuous mapping theorem. Next apply the continuous mapping theorem to this altered form of (9.3) using the continuous mapping

$$h(x_1, x_2) = (x_1/E\{N_1^{(1)}\}, x_2/E\{N_1^{(2)}\})$$

to obtain

(9.5) $$n^{1/2} \begin{bmatrix} (\bar{\tau}/\overline{N}^{(1)}) - r^{(1)} \\ (\bar{\tau}/\overline{N}^{(2)}) - r^{(2)} \end{bmatrix} \Rightarrow N(0, B\Sigma B'),$$

as $n \to \infty$. Note that from (9.5) we could construct a simultaneous confidence interval for $(r^{(1)}, r^{(2)})$. Finally, set

$$\sigma^2 = \sigma_{11}/E\{N_1^{(1)}\}^2 + \sigma_{22}/E\{N_1^{(2)}\}^2 - 2\sigma_{12}/(E\{N_1^{(1)}\}E\{N_1^{(2)}\}).$$

A third application of the continuous mapping theorem yields

(9.6) $$\frac{n^{1/2}\left[\{(\bar{\tau}/\overline{N}^{(1)}) - (\bar{\tau}/N^{(2)})\} - (r^{(1)} - r^{(2)})\right]}{\sigma} \Rightarrow N(0,1)$$

as $n \to \infty$. The c.l.t. of (9.6) can be used to construct a confidence interval for $r^{(1)} - r^{(2)}$, provided that an estimate for the constant σ is available. Using the classical method we can estimate σ from the sequence of observations taken in the n 0-cycles of the process V. This estimate for σ is given in Section 3.10.

A special case of the situation just discussed is when the two types of jobs are the same; here there is only one job type, but we elect to mark two jobs. Let $r^{(1)} = r^{(2)} = r$, $\hat{r}^{(\nu)} = \bar{\tau}/\overline{N}^{(\nu)}$, and $\hat{r} = (\hat{r}^{(1)}, \hat{r}^{(2)})$. Then we can use the "method of multiple estimates" applied to (9.5). For any vector $\beta = (\beta_1, \beta_2)$ with $\beta_1 + \beta_2 = 1$, set $\sigma^{(2)}(\beta) = \beta'(B\Sigma B')\beta$. Then we have

$$\frac{n^{1/2}(\beta'\hat{r} - r)}{\sigma(\beta)} \Rightarrow N(0,1)$$

as $n \to \infty$. Next we select that value, β^*, of β, that minimizes $\sigma^2(\beta)$

subject to $\beta e = 1$, where $e = (1,1)$. It turns out that β^* is given by

$$\beta^* = \{1/(e'(B\Sigma B')^{-1}e)\}(B\Sigma B')^{-1}e$$

and

(9.7) $$\sigma^2(\beta^*) = 1/\{e'(B\Sigma B')^{-1}e\}.$$

Since $\beta = (1,0)$ is one possible value of β, using β^* is guaranteed to yield a variance reduction over that obtained by marking just one job. Again, of course, we must estimate the variance $\sigma^2(\beta^*)$ given in (9.7) from the observations recorded.

Estimation of $P\{R^{(1)} \leq x\} - P\{R^{(1)} \leq x\}$

Finally, we consider the estimation of the quantity $P\{R^{(1)} \leq x\} - P\{R^{(2)} \leq x\}$ for a given value of x. This is the most difficult of the problems we treat. Since the value of x is fixed throughout the discussion, in general we suppress in our notation the dependence on x. Again we form 0-cycles based on the response times for the type 1 marked job. Here, however, when a 0-cycle ends, we do not know whether the response time for the type 2 marked job in progress is less than or equal to x. Thus, with respect to the response times for the type 2 marked job, the 0-cycles used previously do not create the i.i.d. blocks needed to establish a c.l.t.. Indeed, we form new cycles by grouping together a random number of consecutive 0-cycles as follows. Set $t_i = \tau_1 + \ldots + \tau_i$, $i \geq 1$. Then let s_i be the start time of the response time for the type 2 marked job underway at the conclusion of the ith 0-cycle. We assume that $V(0) = 0$ and regard the response time for the type 2 marked job underway at the start of the simulation to be greater than (the fixed) x. We do this so that the start of the simulation corresponds to the beginning of one of the new "super-cycles" we are constructing. Defining a random variable γ according to

$$\gamma = \inf \{i \geq 1 : t_i - s_i > x\},$$

the length of the first super-cycle is simply $\tau_1 + \ldots + \tau_\gamma$ and the number of response times for the type ν marked job that start in this super-cycle is $K_1^{(\nu)} = N_1^{(\nu)} + \ldots + N_\gamma^{(\nu)}$. Successive super-cycles are defined in an analogous fashion. Let $Y_k^{(\nu)}$ be the number of response times terminating in the kth super-cycle that are less than or equal to x for example

$$Y_1^{(\nu)} = \sum_{k=0}^{K_1^{(\nu)}-1} 1_{\{R_k^{(\nu)} \leq x\}}.$$

Observe that by the definition of a super-cycle, the first response time of the type 2 marked job terminating within a super-cycle must be greater than x. Thus the sequence of $Y_k^{(2)}$'s are i.i.d. Of course, the $Y_k^{(1)}$'s are i.i.d. also, as are the $K_k^{(1)}$'s and $K_k^{(2)}$'s. We can now form the bivariate c.l.t. analogous to (9.5), namely

$$n^{1/2} \begin{bmatrix} (\overline{Y}^{(1)}/\overline{K}^{(1)}) - P\{R^{(1)} \leq x\} \\ (\overline{Y}^{(2)}/\overline{K}^{(2)}) - P\{R^{(2)} \leq x\} \end{bmatrix} \Rightarrow N(0, B(x)\Sigma(x)B'(x))$$

as $n \to \infty$. The quantity

$$B(x) = \begin{bmatrix} 1/E\{K_1^{(1)}\} & 0 \\ 0 & 1/E\{K_1^{(2)}\} \end{bmatrix}$$

and $\Sigma(x) = \{\sigma_{ij}(x)\}$ with

$$\sigma_{ij}(x) = E\{[Y_1^{(i)} - K_1^{(i)} P\{R^{(i)} \leq x\}][Y_1^{(j)} - K_1^{(j)} P\{R^{(j)} \leq x\}]\}.$$

Finally, by the same argument that leads to (9.6), we obtain

(9.8) $$n^{1/2} \left[\frac{\overline{Y}^{(1)}}{\overline{K}^{(1)}} - \frac{\overline{Y}^{(2)}}{\overline{K}^{(2)}} - \left(P\{R^{(1)} \leq x\} - P\{R^{(2)} \leq x\}\right) \right] \Big/ \sigma(x) \Rightarrow N(0,1)$$

as $n \to \infty$. The quantity

$$\sigma^2(x) = (\sigma_{11}(x)/(E\{K_1^{(1)}\})^2)$$
$$+ (\sigma_{22}(x)/(E\{K_1^{(2)}\})^2) - 2\sigma_{12}(x)/(E\{n^{(1)}\}E\{K_1^{(2)}\}).$$

We can estimate the quantity $\sigma(x)$ form the observations in the n super-cycles using the classical method; see Section 3.10. Then we construct confidence intervals for $P\{R^{(1)} \leq x\} - P\{R^{(2)} \leq x\}$ from (9.8) in the usual way.

Example 9.9. (Cyclic Queues With Two Job Types) Consider a closed network with two job types and two service centers; see Figure 3.6. There are N jobs in the network: N_1 jobs of type 1 and N_2 jobs of type 2. Upon completion of service in center 1, a type ν job joins the queue at center 1 (with probability $p^{(\nu)}$) or (with probability $1 - p^{(\nu)}$) joins the queue in center 2. Upon completion of service at center 2, jobs join the queue at center 1. At both service centers, type 1 jobs have non-preemptive priority over type 2 jobs. Jobs of the same type at either of the centers receive service in order of their arrival at the center. We assume that all service times are mutually independent; jobs of type ν at center i receive service that is exponentially distributed with parameter $\lambda_i^{(\nu)}$. The (limiting) response time, $R^{(\nu)}$, for type ν jobs that we consider is the time that starts when, upon completion of service at center 2, a type ν job enters the queue at center 1 and terminates when the job next enters the queue at center 1.

Suppose that there are two job classes: class 1 jobs at center 1 and class 2 jobs at center 2. Each center sees both job types, but only one job class. The irreducible routing matrices $P^{(\nu)}$ are of the form

$$P^{(\nu)} = \begin{matrix} p^{(\nu)} & 1 - p^{(\nu)} \\ 1 & 0 \end{matrix}$$

Since type 1 jobs have priority over type 2 jobs at both centers, ordered by decreasing priority the (type, class) pairs served at

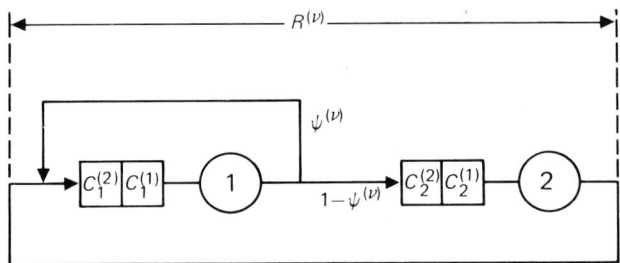

Figure 3.6. Network with multiple job types

center i are $j_1(i) = (1,i)$ and $j_2(i) = (2,i)$, $i = 1,2$. It is sufficient to take as the component $S_i(t)$ in the vector $Z(t)$ the type of job in service at center i at time t, rather than the (type, class) pair. Set

$$Z(t) = (C_1^{(2)}(t), C_1^{(1)}(t), S_1(t), C_2^{(2)}(t), C_2^{(1)}(t), S_2(t)),$$

where $C_i^{(\nu)}(t)$ is the number of type ν jobs in queue at center i at time t and $S_i(t)$ is the type of job in service at center i at time t. ($S_i(t)$ equals zero if center i is idle at time t.) Let $N_\nu(t)$ be the position from the top of the type ν marked job in the linear job stack and set

$$X(t) = (Z(t), N_1(t), N_2(t)).$$

Let $L(t)$ be the last state visited by the augmented job stack process $X = \{X(t): t \geq 0\}$ before jumping to $X(t)$ and set $V(t) = (L(t), X(t))$.

With $N = 2$ jobs, the state space of the augmented job stack process X is

$$G = \{(0,0,0,1,0,1,2,1),(0,0,0,0,1,2,1,2),(0,0,1,0,0,2,1,2)\}$$

$$\cup \{(0,0,2,0,0,1,2,1),(1,0,1,0,0,0,2,1),(0,1,2,0,0,0,1,2)\}.$$

The subsets $A_1^{(1)}$ and $A_2^{(1)}$ of G defining the start of response times for the type 1 marked job are

$$A_1^{(1)} = \{(0,0,0,1,0,1,2,1),(0,0,2,0,0,1,2,1)\}$$

and

$$A_2^{(1)} = \{(0,0,1,2,0,2,1,2),(0,1,2,0,0,0,1,2)\}.$$

Similarly, the subsets $A_1^{(2)}$ and $A_2^{(2)}$ of G defining the start of response times for the type 2 marked job are

$$A_1^{(2)} = \{(0,0,0,0,1,2,1,2),(0,0,1,0,0,2,1,2)\}$$

and

$$A_2^{(2)} = \{(0,0,2,0,0,1,2,1),(1,0,1,0,0,0,2,1)\}.$$

Since $R^{(1)}$ and $R^{(2)}$ are response times, $B_1^{(\nu)} = A_1^{(\nu)}$ and $B_2^{(\nu)} = A_2^{(\nu)}$, $\nu = 1,2$; see Figure 3.7.

It is easy to check that the state space, F, of the process V has nine states. The subsets $S^{(\nu)}$ of F defining the starts of response times for the type ν marked job are

$$S^{(1)} = \{(0,0,0,1,0,1,2,1,0,0,1,0,0,2,1,2)\}$$
$$\cup \{(0,0,2,0,0,1,2,1,0,1,2,0,0,0,1,2)\}$$

and

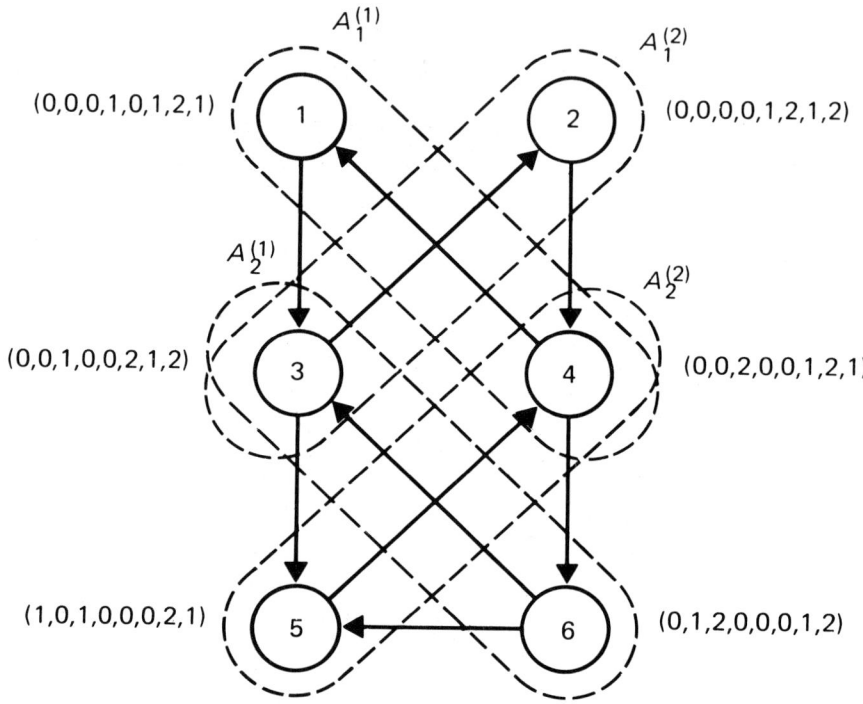

Figure 3.7. State transitions and subsets of G for response times $R^{(1)}$ and $R^{(2)}$

$$S^{(2)} = \{(0,0,0,0,1,2,1,2,0,0,2,0,0,1,2,1)\}$$
$$\cup \{(0,0,1,0,0,2,1,2,1,0,1,0,0,0,2,1)\},$$

respectively; see Figure 3.8. The enumeration of the six states of G is as given in Figure 3.7 so that, e.g., (1,3) denotes the state $(0,0,0,1,0,1,2,1,0,0,1,0,0,2,1,2) \in F$.

Table 3.5 Simulation Results for Response Times in Cyclic Queues With Two Job Types.

$N_1 = 1, N_2 = 1, p = 0.75, \lambda_1 = 1, \lambda_2 = 0.5.$

Return State is (0,0,2,0,0,1,2,1,0,1,2,0,0,0,1,2)

	\multicolumn{5}{c}{No. of Cycles for Type 1 Marked Job}				
	100	200	400	800	1000
Simulated time	903.00	1959.09	4134.11	8211.57	10172.83
No. of transitions/ cycle (CTMC)	4.46	4.85	5.18	5.12	5.15
No. of type 1 response times/cycle	1.30	1.40	1.48	1.46	1.47
No. of type 2 response times/cycle	0.56	0.65	0.75	0.74	0.75
$E\{R^{(1)}\}$ (7)	6.946 ±0.6334	6.997 ±0..3931	6.995 ±0.2703	7.018 ±0.2033	6.920 ±0.1800
$E\{R^{(2)}\}$ (14)	16.125 ±2.8713	15.187 1.8017	13.780 ±0.9851	13.965 ±0.7361	13.729 ±0.6305
$E\{R^{(1)}\} - E\{R^{(2)}\}$ (−7)	−7.179 ±2.7421	−8.190 ±1.7316	−6.785 ±0.9145	−6.947 ±0.6809	−6.808 ±0.5848

Simulation results for $N = 2$ jobs with $p^{(1)} = p^{(2)} = p = 0.75$, $\lambda_1^{(1)} = \lambda_1^{(2)} = \lambda_1 = 1$ and $\lambda_2^{(1)} = \lambda_2^{(2)} = \lambda_2 = 0.5$ are given in Tables 3.5-3.8. There is one type 1 job and one type 2 job. The routing and service requirements of the two job types are the same; the two jobs differ only with respect to the non-preemptive priority given (at each center) to the type 1 job. Exponential service times have been generated by logarithmic transformation of uniform random numbers obtained as in Section 3.5. Independent streams of exponential random numbers (resulting from different seeds of the uniform random number generator) have been used to generate individual exponential holding time sequences.

Table 3.6 Estimates for Percentiles of Type 1 Response Times in Cyclic Queues With Two Job Types.

$N_1 = 1, N_2 = 1, p = 0.75, \lambda_1 = 1, \lambda_2 = 0.5.$

Return State is (0,0,2,0,0,1,2,1,0,1,2,0,0,0,1,2)

	No. of Cycles for Type 1 Marked Job				
	100	200	400	800	1000
$P\{R^{(1)} \leq 4\}$	0.2384 ±0.0622	0.2536 ±0.0417	0.2555 ±0.0301	0.2641 ±0.0217	0.2639 ±0.0192
$P\{R^{(1)} \leq 8\}$	0.6692 ±0.0683	0.6714 ±0.0444	0.6717 ±0.0308	0.6709 ±0.0221	0.6802 ±0.0201
$P\{R^{(1)} \leq 12\}$	0.8923 ±0.0422	0.8786 ±0.0293	0.8832 ±0.0205	0.8769 ±0.0159	0.8830 ±0.0140
$P\{R^{(1)} \leq 16\}$	0.9461 ±0.0311	0.9536 ±0.0198	0.9594 ±0.0135	0.9547 ±0.0105	0.9605 ±0.0088
$P\{R^{(1)} \leq 20\}$	0.9923 ±0.0127	0.9892 ±0.0100	0.9915 ±0.0061	0.9880 ±0.0052	0.9898 ±0.0043

In Tables 3.5-3.8, the return state defining 0-cycles of the response time for the type 1 job is the state (0,0,2,0,0,1,2,1,0,1,2,0,0,0,1,2). This corresponds to a response time for the type 1 (marked) job starting when the type 2 (marked) job is in service at center 1. Table 3.5 summarizes results of the simulation and gives point estimates for the quantities $E\{R^{(1)}\}$, $E\{R^{(2)}\}$ and $E\{R^{(1)}\} - E\{R^{(2)}\}$ over a range of number of cycles of the type 1 marked job. Theoretical values for these quantities are shown in parentheses. Thus, for example, 100 cycles of the type 1 marked job were observed in the simulated time interval (0,903.00) and there were a total of 446 transitions in the CTMC $\{X(t):t \geq 0\}$. A total of 130 response times for the type 1 (marked) job were observed along with 56 response times for the type 2 (marked job). For the quantity $E\{R^{(1)}\} = 7$, the point estimate 6.946 was obtained, and the 90% confidence interval had half length 0.6334. Note that for $E\{R^{(1)}\}$ and $E\{R^{(2)}\}$, all of the confidence intervals contain the theoretical values. In the case of $E\{R^{(1)}\} - E\{R^{(2)}\}$, the confidence intervals based on (9.6) also contain the theoretical value. Table 3.6 gives point estimates obtained for $P\{R^{(1)} \leq x\}$, with $x = 4, 8, 12, 16$, and 20.

Table 3.7 gives, for the several values of x, point estimates for $P\{R^{(1)} \leq x\} - P\{R^{(2)} \leq x\}$, based on the use of super-cycles and (9.8). Thus, for $x = 4$, 100 cycles based on response times for the type 1 job resulted in 37 super-cycles defined by response times for

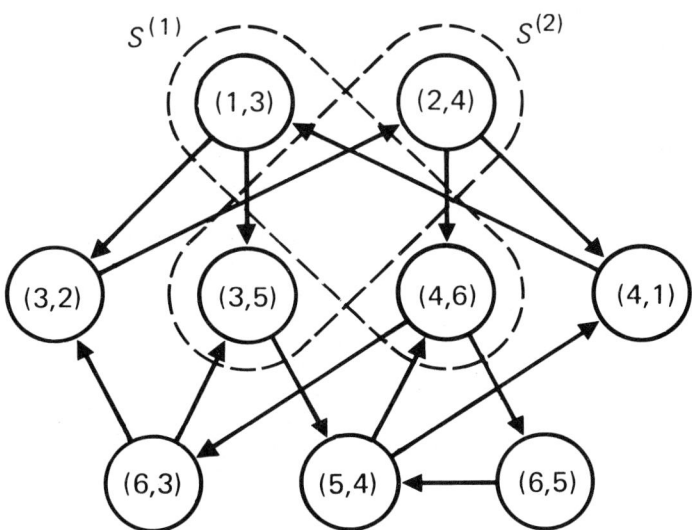

Figure 3.8. State transitions in Markov chain V

the type 2 job greater than x. Note that the number of cycles for the type 1 marked job has been fixed, and for each x the estimates for $P\{R^{(1)} \leq x\} - P\{R^{(2)} \leq x\}$ computed from the resulting random number of super-cycles. Table 3.8 gives estimates of the quantities $P\{R^{(2)} \leq x\}$ obtained from the standard regenerative method applied to these super-cycles. An overall observation is that the lengths of confidence intervals obtained for $P\{R^{(1)} \leq x\}$ and $P\{R^{(2)} \leq x\}$ are roughly comparable.

Table 3.7 Estimates for Difference of Percentiles of Response Times in Cyclic Queues With Two Job Types

$$N_1 = 1, N_2 = 1, p = 0.75, \lambda_1 = 1, \lambda_2 = 0.5.$$

Return State is (0,0,2,0,0,1,2,1,0,1,2,0,0,0,1,2)

	No. of Cycles for Type 1 Marked Job				
	100	200	400	800	1000
$P\{R^{(1)} \leq 4\} - P\{R^{(2)} \leq 4\}$	0.1254 ±0.0784	0.1295 ±0.0627	0.1342 ±0.0417	0.1332 ±0.0275	0.1384 ±0.0192
No. of super-cycles	37	79	181	347	438
$P\{R^{(1)} \leq 8\} - P\{R^{(2)} \leq 8\}$	0.3988 ±0.1227	0.3409 ±0.0817	0.3111 ±0.0525	0.3036 ±0.0373	0.3125 ±0.0201
No. of super-cycles	29	61	134	254	319
$P\{R^{(1)} \leq 12\} - P\{R^{(2)} \leq 12\}$	0.3915 ±0.1424	0.3543 ±0.1112	0.3331 ±0.0677	0.3259 ±0.0496	0.3280 ±0.0410
No. of super-cycles	22	46	93	180	224
$P\{R^{(1)} \leq 16\} - P\{R^{(2)} \leq 16\}$	0.2850 ±0.1288	0.2970 ±0.1092	0.2693 ±0.0606	0.2693 ±0.0489	0.2636 ±0.0388
No. of super-cycles	15	32	65	129	153
$P\{R^{(1)} \leq 20\} - P\{R^{(2)} \leq 20\}$	0.2422 ±0.1081	0.2470 ±0.0825	0.2119 ±0.0530	0.2078 ±0.0415	0.2009 ±0.0343
No. of super-cycles	11	24	43	84	104

We have concentrated on problems associated with the estimation of characteristics of response times for the two types of jobs. The estimation of characteristics of two passage times, or one response time and one passage time, is in general easier. This is because there is the possibility of forming, from 0-cycles based on one type of job, super-cycles that terminate when no passage time of the other type of job is underway.

Only the case of two job types has been considered explicitly. The estimation methods of this section apply equally well to networks having more than two job types. The state space that results from the augmentation of the vector $X(t)$ (by components to track a marked job of each of the job types) is of course larger.

Table 3.8 Estimates for Percentiles of Type 2 Response Times in Cyclic Queues With Two Job Types.

$$N_1 = 1, N_2 = 1, p = 0.75, \lambda_1 = 1, \lambda_2 = 0.5.$$

Return State is (0,0,2,0,0,1,2,1,0,1,2,0,0,0,1,2)

	\multicolumn{5}{c}{No. of Cycles for Type 1 Marked Job}			
	100	200	400	800	1000
$P\{R^{(2)} \leq 4\}$	0.1071	0.1240	0.1200	0.1310	0.1255
	±0.0852	±0.0583	±0.0312	±0.0209	±0.0183
No. of super-cycles	37	79	181	347	438
$P\{R^{(2)} \leq 8\}$	0.2679	0.3281	0.3600	0.3673	0.3681
	±0.0963	±0.0751	±0.0475	±0.0323	±0.0280
No. of super-cycles	29	61	134	254	319
$P\{R^{(2)} \leq 12\}$	0.5000	0.5234	0.5500	0.5510	0.5548
	±0.1118	±0.0799	±0.0474	±0.0336	±0.0286
No. of super-cycles	22	46	93	180	224
$P\{R^{(2)} \leq 16\}$	0.6607	0.6563	0.6900	0.6854	0.6969
	±0.1115	±0.0687	±0.0419	±0.0316	±0.0268
No. of super-cycles	15	32	65	129	153
$P\{R^{(2)} \leq 20\}$	0.7500	0.7422	0.7793	0.7802	0.7889
	±0.0986	±0.0582	±0.0405	±0.0284	±0.0243
No. of super-cycles	11	24	43	84	104

3.10 Estimation of Variance Constants

We first consider estimation of the variance constant σ^2 in (9.6) that leads to a confidence interval for $E\{R^{(1)}\} - E\{R^{(2)}\}$. Based on n cycles, compute $\hat{\sigma}_{11}$ as an estimate of

$$\sigma_{ii} = E\{(Z_k^{(i)})^2\} = \text{var}(\tau_k) - 2r^{(i)} \text{cov}(\tau_k, N_k^{(i)}) + (r^{(i)})^2 \text{var}(N_k^{(i)})$$

according to

$$\hat{\sigma}_{ii} = s_{11} - 2\hat{r}^{(i)} s_{12}^{(i)} + (\hat{r}^{(i)})^2 s_{22}^{(i)},$$

where

$$s_{11} = \frac{1}{n-1} \sum_{j=1}^{n} (\tau_j - \bar{\tau})^2,$$

$$s_{12}^{(i)} = \frac{1}{n-1} \sum_{j=1}^{n} (\tau_j - \bar{\tau})(N_j^{(i)} - \bar{N}^{(i)}),$$

and

$$s_{22}^{(i)} = \frac{1}{n-1} \sum_{j=1}^{n} (N_j^{(i)} - \bar{N}^{(i)})^2.$$

The quantities $\bar{\tau}$ and $\bar{N}^{(i)}$ are sample means over the n cycles and $\hat{r}^{(i)} = \bar{\tau}/\bar{N}^{(i)}$. Finally, compute $\hat{\sigma}_{12}$ as an estimate of

$$\sigma_{12} = \text{var}(\tau_k) - r^{(1)} \text{cov}(\tau_k, N_k^{(1)}) - r^{(2)} \text{cov}(\tau_k, N_k^{(2)})$$

$$+ r^{(1)} r^{(2)} \text{cov}(N_k^{(1)}, N_k^{(2)})$$

according to

$$\sigma_{12} = s_{11} - \hat{r}^{(1)} s_{12}^{(1)} - \hat{r}^{(2)} s_{12}^{(2)} + \hat{r}^{(1)} \hat{r}^{(2)} s_{22},$$

where s_{11}, $s_{12}^{(1)}$, and $s_{12}^{(2)}$ are as before, and

$$s_{22} = \frac{1}{n-1} \sum_{j=1}^{n} (N_j^{(1)} - \overline{N}^{(1)})(N_j^{(2)} - \overline{N}^{(2)}).$$

Then estimate σ^2 according to

$$\hat{\sigma}^2 = \frac{\hat{\sigma}_{11}}{(\overline{N}^{(1)})^2} + \frac{\hat{\sigma}_{22}}{(\overline{N}^{(2)})^2} - \frac{2\hat{\sigma}_{12}}{\overline{N}^{(1)}\overline{N}^{(2)}}.$$

In an analogous manner, we estimate the variance constant $\sigma^2(x)$ appearing in (9.8) that leads to a confidence interval for $P\{R^{(1)} \leq x\} - P\{R^{(2)} \leq x\}$. Based on n super-cycles, compute $\hat{\sigma}_{ii}(x)$ as an estimate of

$$\hat{\sigma}_{ii}(x) = \mathrm{var}(Y_k^{(i)}) - 2P\{R^{(i)} \leq x\} \mathrm{cov}(Y_k, K_k^{(i)})$$

$$+ (P\{R^{(i)} \leq x\})^2 \mathrm{var}(K_k^{(i)})$$

according to

$$\hat{\sigma}_{ii}(x) = s_{11}^{(i)}(x) - 2\left(\frac{\overline{Y}^{(i)}}{\overline{K}^{(i)}}\right) s_{12}^{(i)}(x) + \left(\frac{\overline{Y}^{(i)}}{\overline{K}^{(i)}}\right)^2 s_{22}^{(i)}(x),$$

where

$$s_{11}^{(i)}(x) = \frac{1}{n-1} \sum_{j=1}^{n} (Y_j^{(i)} - \overline{Y}^{(i)})^2,$$

$$s_{12}^{(i)}(x) = \frac{1}{n-1} \sum_{j=1}^{n} (Y_j^{(i)} - \overline{Y}^{(i)})(K_j^{(i)} - \overline{K}^{(i)}),$$

and

$$s_{22}^{(i)}(x) = \frac{1}{n-1} \sum_{j=1}^{n} (K_j^{(i)} - \overline{K}^{(i)})^2.$$

The quantities $\bar{Y}^{(i)}$ and $\bar{K}^{(i)}$ are sample means over the n super-cycles. Finally, compute $\hat{\sigma}_{12}(x)$ as an estimate of

$$\sigma_{12}(x) = \text{cov}(Y_k^{(1)}, Y_k^{(2)}) - P\{R^{(1)} \le x\} \text{cov}(Y_k^{(2)}, K_k^{(1)})$$

$$- P\{R^{(2)} \le x\} \text{cov}(Y_k^{(1)}, K_k^{(2)})$$

$$+ P\{R^{(1)} \le x\} P\{R^{(2)} \le x\} \text{cov}(K_k^{(1)}, K_k^{(2)})$$

according to

$$\sigma_{12}(x) = s_{11}^{(1)}(x) - \left(\frac{\bar{Y}^{(1)}}{\bar{K}^{(1)}}\right) s_{12}^{(1)}(x) - \left(\frac{\bar{Y}^{(2)}}{\bar{K}^{(2)}}\right) s_{21}^{(2)}(x)$$

$$+ \left(\frac{\bar{Y}^{(1)} \bar{Y}^{(2)}}{\bar{K}^{(1)} \bar{K}^{(2)}}\right) s_{22}(x),$$

where $s_{11}^{(1)}(x)$, $s_{22}^{(1)}(x)$, and $s_{12}^{(2)}(x)$ are as before, and

$$s_{22}(x) = \frac{1}{n-1} \sum_{j=1}^{n} (K_j^{(1)} - \bar{K}^{(1)})(K_j^{(2)} - \bar{K}^{(2)}).$$

Then estimate $\sigma^2(x)$ according to

$$\hat{\sigma}^2(x) = \frac{\hat{\sigma}_{11}(x)}{(\bar{K}^{(1)})^2} + \frac{\hat{\sigma}_{22}(x)}{(\bar{K}^{(2)})^2} - \frac{2\hat{\sigma}_{12}(x)}{\bar{K}^{(1)} \bar{K}^{(2)}}.$$

Chapter 4

Non-Markovian Networks of Queues

In Chapter 3 we developed regenerative simulation methods for networks of queues with exponential service times and Markovian job routing. In this chapter, we derive analogous estimation procedures for networks with general service times. The development requires that we restrict attention to networks that have a "single state" in which all jobs are at the same service center with exactly one job in service. Regenerative cycles are defined in terms of entrances to the single state.

4.1 Networks with Single States

As in Chapter 3, we consider closed networks of queues having a finite number of *jobs* (customers), N, a finite number of *service centers*, s, and a finite number of (mutually exclusive) *job classes*, c. At every epoch of continuous time each job is in exactly one job class, but jobs may change class as they traverse the network. Upon completion of service at center i a job of class j goes to center k and changes to class l with probability $p_{ij,kl}$, where

$$P = \{p_{ij,kl}: (i,j),(k,l) \in C\}$$

is a given irreducible stochastic matrix and $C \subseteq \{1,2,...,s\} \times \{1,2,...,c\}$ is the set of (center, class) pairs in the network. In accordance with the routing matrix P, some centers may never see jobs of certain classes. At each service center jobs queue and receive service according to a fixed priority scheme among classes; the priority scheme may differ from center to center. Within a class at a center, jobs receive service according to a fixed queue service discipline. According to a fixed procedure for each center, a job in service may or may not be preempted when a job of higher priority joins the queue at the center. (We assume that any interruption of service is of the preemptive-repeat type.) A job that has been preempted receives additional service at the center before any other job of its class at the center receives service.

We assume that service times are mutually independent and that the distribution of service times at a center has finite first and second moments and a density function that is continuous and positive on $(0, +\infty)$. Parameters of the service time distribution may depend on the service center, the class of job being served, and the "state" (as defined below) of the entire network.

Let $S_i(t)$ be the class of the job receiving service at center i at time t, where $i = 1,2,...,s$; by convention $S_i(t) = 0$ if at time t there are no jobs at center i. If center i has more than one server, we take $S_i(t)$ to be a vector that records the class of the job receiving service from each server at center i. (Specifically, we enumerate the servers at center i as $1,2,...,s(i)$ and set

$$S_i(t) = (S_{i,1}(t), S_{i,2}(t),...,S_{i,s(i)}(t)),$$

where $S_{i,m}(t)$ is the class of the job receiving service from server m at center i at time t.) The job classes served at center i ordered by decreasing priority are $j_1(i), j_2(i),...,j_{k(i)}(i)$, elements of the set $\{1,2,...,c\}$. We denote by $C^{(i)}_{j_1}(t),...,C^{(i)}_{j_{k(i)}}(t)$ the number of jobs in queue at time t of the classes of jobs serviced at center i.

4.1 Networks with Single States

As before, we order the N jobs in a linear stack and define the state of the system at time t to be the vector $Z(t)$ given by

(1.1) $\quad Z(t) = \left(C_{j_{k(1)}}^{(1)}(t),\ldots,C_{j_1}^{(1)}(t),S_1(t);\ldots;C_{j_{k(s)}}^{(s)}(t),\ldots,C_{j_1}^{(s)}(t),S_s(t) \right).$

The *job stack at time t* then corresponds to the nonzero components in the vector $Z(t)$ and thus is an ordering of the jobs by class at the individual centers. Within a class at a particular service center, jobs appear in the job stack in order of their arrival at the center, the latest to arrive being closest to the top of the stack. (A job that has been preempted appears at the head of its job class queue.) The process $Z = \{Z(t):t \geq 0\}$ has a finite state space, D^*, and is called the *job stack process*.

Let $U(z)$ be the set of all (center, class) pairs $(i,j) \in C$ such that there is a job of class j in service at center i when the job stack process is in state z. For $z,z' \in D^*$ and $u = (i,j) \in U(z)$, let $q(z';z,u(m))$ be the probability that the job stack process Z makes a transition to state z' given that in state z there is a completion of service to a job of class j by server m at center i. (We write $z \to z'$ if $q(z';z,u(m)) > 0$.) For $z,z' \in D^*$, we say that z' *is accessible from* z and write $z \sim z'$ if there exists a finite sequence $z_1,z_2,\ldots,z_n$ of states of the job stack process, (center, class) pairs $u_{i_0},u_{i_1},\ldots,u_{i_n}$, and servers $m_0,m_1,\ldots,m_n$ such that

$$q(z_1;z,u_{i_0}(m_0))q(z_2;z_1;u_{i_1}(m_1))\cdots q(z';z_n,u_{i_n}(m_n)) > 0.$$

When $z \sim z'$ and $z' \sim z$ we say that z and z' *communicate*.

We assume throughout that there exists a "target state", z^*, of the job stack process such that $z \sim z^*$ for all $z \in D^*$ so that the set

(1.2) $\quad\quad\quad\quad D = \{z \in D^* : z^* \sim z\}$

of all states accessible from z^* is *irreducible* in the sense that z and z' communicate for all $z,z' \in D$. (To see this, let $z_1,z_2 \in D$. The definition of z^* ensures that $z_1 \sim z^*$ and $z^* \sim z_2$ because $z_2 \in D$. Therefore, $z_1 \sim z_2$.)

Proposition 1.3 provides conditions on the building blocks of a network of queues which ensure the existence of a target state of the job stack process. It is sufficient that for some service center, i_0, either $k(i_0) = 1$ or service at center i_0 to a job of class $j_{k(i_0)}$ (the lowest priority job class seen by center i_0) is preempted when any other job of higher priority joins the queue. Then state, $z^*_{i_0}$, in which there is one job of class $j_{k(i_0)}(i_0)$ in service at center i_0 and $N - 1$ jobs of class $j_{k(i_0)}(i_0)$ in queue at center i_0 (or in service if center i_0 is a multiple server center), is a target state of the job stack process and the set $D = \{z \in D^* : z^*_{i_0} \sim z\}$ is irreducible. The proof is constructive; cf. Proposition 1.9 of Chapter 3.

Proposition 1.3. Suppose that the routing matrix P is irreducible and that for some service center, i_0, either $k(i_0) = 1$ or service to a job of class $j_{k(i_0)}(i_0)$ at center i_0 is preempted when any other job of higher priority joins the queue. Then z and z' communicate for all $z, z' \in D$.

As there are priorities among job classes, the set $D^* - D$ may be nonempty even if the routing matrix P is irreducible. For example, recall the system of cyclic queues with preemptive priority and the job stack process defined by (1.3) of Chapter 3. For $N = 2$ jobs, it is easy to check that state $(0,0,1,2)$ is not accessible from any state of the job stack process; consequently $(0,0,1,2) \in D^* - D$. The remaining five states of the process comprise the set D and all pairs of states in D communicate.

We restrict attention to job stack processes that have a "single state" in which all jobs are at the same service center with exactly one job in service. Regenerative cycles are defined in terms of entrances to the single state. Let h be the function having domain $D \times \{1, 2, \ldots, N\}$ and taking values in the set C such that $h(z, n) = (i, j)$ when the job in position n of the job stack associated with $z \in D$ is of class j at center i.

Definition 1.4. An element $z_0 \in D$ is called a *single state of the job stack process* if $h(z_0,n) = (i_0, j_{l_n}(i_0))$ for some single server center, i_0, $n = 1,2,...,N$.

According to this definition, a single state of the job stack process corresponds to a configuration of the job stack such that all jobs are at the same center (i_0) with exactly one job in service. (Recall that $j_1(i_0), j_2(i_0),...,j_{k(i_0)}(i_0)$ are the classes of jobs served at center i_0. Definition 1.4 requires that the class, $j_{l_n}(i_0)$, at center i_0 of the job in position n satisfy $j_{l_n}(i_0) \in \{j_1(i_0), j_2(i_0),...,j_{k(i_0)}(i_0)\}$ for all n.) We assume that a single state of the job stack process exists. In addition, we assume that the initial state of the job stack process is a single state. Note that state $z_{i_0}^*$ is a single state of the job stack process if i_0 is a single server service center that satisfies the conditions of Proposition 1.3.

4.2 Regenerative Simulation of Non-Markovian Networks

The key result of this section is that the job stack process is a regenerative process in continuous time and the expected time between regeneration points is finite. The argument rests on representation of the job stack process restricted to the set D as an irreducible GSMP with unit speeds. (There are no zero speeds since all service interruptions are of the preemptive-repeat type.)

Proposition 2.1 states a recurrence result for irreducible, finite state GSMP's with unit speeds. Recall that a GSMP having state space, S, event set, E, and unit speeds is irreducible if for each pair $s, s' \in S$ there exists a finite sequence of states $s_1, s_2,...,s_n \in S$ and events $e_{i_0}, e_{i_1},...,e_{i_n} \in E$ such that

$$p(s_1; s, e_{i_0}) p(s_2; s_1, e_{i_1}) \cdots p(s'; s_n, e_{i_n}) > 0.$$

Proposition 2.1. Let $\{X(t): t \geq 0\}$ be an irreducible GSMP with a finite state space, S, event set, E, and unit speeds. Denote by ζ_n the nth time at which the GSMP makes a state transition, $n \geq 0$.

Suppose that for all $s, s' \in S$, $e^* \in E$, and $e' \in N(s'; s, e^*)$ the clock setting distribution $F(\cdot; s', e', s, e^*)$ has a finite mean and a density function that is continuous and positive on $(0, +\infty)$. Then $P\{X(\zeta_n) = s_0 \text{ i.o.}\} = 1$ for all $s_0 \in S$.

The idea of the proof is to associate the vector, $C(s_0)$, of clock readings with a state, s_0, of the GSMP. It is sufficient to show that the GSSMC associated with state transitions of the GSMP returns infinitely often to the set $\{s_0\} \times C(s_0)$; it follows immediately that $P\{X(\zeta_n) = s_0 \text{ i.o.}\} = 1$. The GSSMC returns infinitely often to the set $\{s_0\} \times C(s_0)$ provided that

(i) the GSMP is irreducible;
(ii) each clock setting distribution has finite mean and a density function that is continuous and positive on $(0, +\infty)$; and
(iii) a "recurrence measure" assigns positive measure to the set $\{s\} \times C(s)$ for all states s of the process.

The GSMP associated with the job stack process $Z = \{Z(t): t \geq 0\}$ has state space, D, and event set, E, where event $e =$ "service completion to a job of class j by server m at center i" is active in state z ($e \in E(z)$) if and only if (center, class) pair $(i, j) \in U(z)$ and in state z a job of class j the class of job receives service from server m at center i. Jumps of the process are governed by the probability mass function $q(\cdot; z, e)$. At a jump from state z to state z' triggered by event e, new clock values are generated for each event $e' \in E(z') - (E(z) - \{e\})$. The distribution of such a new clock value (a service time for a job class at some center) is denoted by $F(\cdot; z', e', z, e)$; by assumption, each has finite mean and a density function that is continuous and positive on $(0, +\infty)$. For $e' \in E(z') \cap (E(z) - \{e\})$, the old clock reading is kept after the jump. For $e' \in (E(z) - \{e\}) - E(z')$, event e' ceases to be scheduled after the jump. Denote the set of all possible events that can occur by $E = \{e_1, e_2, \ldots, e_M\}$ and observe that if all centers are single server centers there is a one-to-one correspondence between E and C, the set of (center, class) pairs in the network. With each $z \in D$, associate the set of clock readings

$$C(z) = \{(c_1, c_2, \ldots, c_M) : c_i \geq 0 \text{ and } c_i > 0 \text{ if and only if } e_i \in E(z)\},$$

where c_i is the reading on the clock corresponding to event e_i.

Proposition 2.2. Let $M(t)$ be the last state occupied by the job stack process $Z = \{Z(t) : t \geq 0\}$ before jumping to the state occupied at time t and set $Y(t) = (M(t), Z(t))$. Denote by ζ_n the nth time at which the process $Y = \{Y(t) : t \geq 0\}$ makes a state transition, $n \geq 0$. Suppose that the routing matrix P is irreducible and for some service center, i_0, either $k(i_0) = 1$ or service to a job of class $j_{k(i_0)}(i_0)$ at center i_0 is preempted when any other job of higher priority joins the queue. Then

(2.3) $$P\{Y(\zeta_n) = (z, z') \text{ i.o.}\} = 1$$

for all $z, z' \in D$ such that $z \to z'$.

Proof. The process Y is a GSMP with finite state space $\{(z, z') : z, z' \in D \text{ and } z \to z'\}$ and the event $e = $ "service completion to a job of class j by server m at center i" is active in state (z, z') if and only if $(i, j) \in U(z')$ and the class of job receiving service from server m at center i is j. Proposition 2.1 ensures that (2.3) holds provided that (i) the GSMP Y is irreducible in the sense that y and y' communicate for all states y, y' of the process and (ii) each clock setting distribution has finite mean and a density function that is continuous and positive on $(0, +\infty)$. Condition (i) holds as a consequence of Proposition 1.3; $z_2 \sim z_1'$ implies $y \sim y'$ for $y = (z_1, z_2)$ and $y' = (z_1', z_2')$. Condition (ii) holds by assumption. □

Now select a single state, z_0, of the job stack process Z and a state, z_0', such that $z_0 \to z_0'$. By Proposition 2.2 we know that $P\{Y(\zeta_n) = (z_0, z_0') \text{ i.o.}\} = 1$. Furthermore at such a time, ζ_n, the only (two) clocks that are active have just been set. Since the jumps of the process Z from state z_0 are governed by the Markovian mass function $q(\cdot;z,e)$ and the only active clocks were set at time ζ_n, the future evolution of the process Z is independent of the history of the process before time ζ_n and has the same distribution as it does when $Z(0) = z_0'$. Thus, the subsequence of state transition times, ζ_n, at which $Y(\zeta_n) = (z_0, z_0')$ are regeneration points for the process Z. Since the state space, D, of the process is finite and the clock setting distributions have finite mean, the expected time between regeneration points is finite.

Proposition 2.4. The job stack process $Z = \{Z(t): t \geq 0\}$ is a regenerative process in continuous time and the expected time between regeneration points is finite.

Since the state space of the job stack process is discrete and the expected time between regeneration points is finite, $Z(t) \Rightarrow Z$ as $t \to \infty$ by Proposition 2.3 of Chapter 2.

Let f be a real-valued function with domain, D, and suppose that the goal of the simulation is the estimation of $r(f) = E\{f(Z)\}$. To obtain estimates for $r(f)$, begin the simulation of the job stack process Z with $Z(0) = z_0'$. Carry out the simulation in cycles defined by the successive times that $\{Y(\zeta_n): n \geq 0\}$ makes a transition to state (z_0, z_0'). Let τ_m be the length of the mth cycle, $m \geq 1$. Set $T_0 = 0$ and

$$T_m = \tau_1 + \ldots + \tau_m.$$

Also set

$$Y_m(f) = \int_{T_{m-1}}^{T_m} f(Z(u))\,du.$$

Proposition 2.5. The sequence of pairs of random variables $\{(Y_m(f), \tau_m): m \geq 1\}$ are independent and identically distributed.

The general result for regenerative processes establishes a ratio formula for $r(f)$.

Proposition 2.6. Provided that $E\{|f(Z)|\} < \infty$,

$$E\{f(Z)\} = \frac{E\{Y_1(f)\}}{E\{\tau_1\}}.$$

With these results

(2.7) $$\hat{r}(n) = \frac{\overline{Y}(n)}{\overline{\tau}(n)} = \frac{\sum_{m=1}^{n} Y_m(f)}{\sum_{m=1}^{n} \tau_m}$$

is a strongly consistent point estimate for $r(f)$ and an asymptotic $100(1 - 2\gamma)\%$ confidence interval is

(2.8) $$\left[\hat{r}(n) - \frac{z_{1-\gamma}\, s(n)}{\overline{\tau}(n)\, n^{1/2}},\, \hat{r}(n) + \frac{z_{1-\gamma}\, s(n)}{\overline{\tau}(n)\, n^{1/2}}\right].$$

The quantity $s^2(n)$ is a strongly consistent point estimate for $\sigma^2(f) = \text{var}\,(Y_1(f) - r(f)\tau_1)$. Using Wald's second moment identity and the fact that the service time distributions have finite second moment, it can be shown that the time between regeneration points has finite second moment. It follows, since the state space of the job stack process is finite, that the variance constant $\sigma^2(f) < \infty$. Asymptotic confidence intervals for $r(f)$ are based on the c.l.t.

(2.9) $$\frac{n^{1/2}\,\{\hat{r}(n) - r(f)\}}{\sigma(f)/E\{\tau_1\}} \Rightarrow N(0,1)$$

as $n \to \infty$.

Algorithm 2.10. (Regenerative Method for Non-Markovian Networks)

1. Select a single state, z_0, of the job stack process and a state, z_0', such that $z_0 \to z_0'$.

2. Set $Z(0) = z_0'$ and simulate the job stack process. Observe a fixed number, n, of cycles defined by the successive times at which $\{Y(\zeta_n): n \geq 0\}$ makes a transition to state (z_0, z_0').

3. Compute the length, τ_m, of the mth cycle and the quantity

$$Y_m(f) = \int_{T_{m-1}}^{T_m} f(Z(u))du,$$

where $T_0 = 0$ and $T_m = \tau_1 + \ldots + \tau_m$.

4. Form the point estimate

$$\hat{r}(n) = \frac{\bar{Y}(n)}{\bar{\tau}(n)}.$$

5. Form the asymptotic $100(1 - 2\gamma)\%$ confidence interval

$$\left[\hat{r}(n) - \frac{z_{1-\gamma} \, s(n)}{\bar{\tau}(n) \, n^{1/2}}, \hat{r}(n) + \frac{z_{1-\gamma} \, s(n)}{\bar{\tau}(n) \, n^{1/2}}\right].$$

4.3 Single States for Passage Times

To obtain estimates for passage times in non-Markovian networks, based on representation as an irreducible GSMP with unit speeds, we show that the augmented job stack process restricted to an appropriate subset of its state space is a regenerative process in continuous time. The choice of a particular sequence of regeneration points leads to an estimation procedure based on one sample path and measurement of passage times for an arbitrarily chosen, distiguished job. This marked job method provides strongly consistent point estimates and asymptotic confidence intervals for general characteristics of limiting passage times.

The "labelled jobs method" in Section 4.9 provides estimates for passage times which correspond to passage through a subnetwork of a given network of queues. With this estimation procedure passage times for all the jobs are recorded by observing a "fully augmented job stack process" that maintains the position of each of the jobs in the job stack. Under a mild restriction on the priorities among job classes, the job stack process observed at the epochs at which passage times terminate is a regenerative process in discrete time. As a consequence, point and interval estimates for characteristics of limiting passage times can be obtained from a single simulation run. Terminations of passage times with no other passage times underway and exactly one job in service are regeneration points for the job stack process observed at termination times. In order for such epochs to exist we must exclude passage times that always terminate with two or more jobs in service. A mild restriction on the priorities among job classes ensures that infinitely many such epochs occur.

We consider closed networks of queues having a finite number of *jobs* (customers), N, a finite number of *service centers*, s, and a finite number of (mutually exclusive) *job classes*, c, as in Section 4.1. Order the N jobs in a linear stack and define the vector $Z(t)$ as

(3.1) $\quad Z(t) = \bigl(C^{(1)}_{j_{k(1)}}(t),...,C^{(1)}_{j_1}(t),S_1(t);...;C^{(s)}_{j_{k(s)}}(t),...,C^{(s)}_{j_1}(t),S_s(t) \bigr).$

(The job stack at time t corresponds to the nonzero components in the vector $Z(t)$ and thus is an ordering of the jobs by class at the individual centers. Within a class at a particular service center, jobs appear in the job stack in order of their arrival at the center, the latest to arrive being closest to the top of the stack. A job that has been preempted appears at the head of its job class queue.) Let $N(t)$ be the position in the job stack at time t of an arbitrarily chosen marked job. Then set

(3.2) $\qquad\qquad\qquad X(t) = (Z(t),N(t)).$

The process $X = \{X(t):t \geq 0\}$ is the *augmented job stack process*. As in Chapter 3, passage times are specified in terms of the marked job by means of four nonempty subsets $(A_1, A_2, B_1,$ and $B_2)$ of the state space, G^*, of the process X. The sets A_1, A_2 [resp., B_1, B_2] jointly define the random times at which passage times for the marked job start [resp., terminate]. The sets A_1, A_2, B_1, and B_2 in effect determine when to start and stop the clock measuring a particular passage time of the marked job. We assume that the start and termination times for the specified passage time strictly alternate:

if $X(\zeta_{n-1}) \in A_1$, $X(\zeta_n) \in A_2$, $X(\zeta_{n-1+k}) \in A_1$, and $X(\zeta_{n+k}) \in A_2$
then $X(\zeta_{n-1+m}) \in B_1$ and $X(\zeta_{n+m}) \in B_2$ for some $0 < m \leq k$;

and

if $X(\zeta_{n-1}) \in B_1$, $X(\zeta_n) \in B_2$, $X(\zeta_{n-1+k}) \in B_1$, and $X(\zeta_{n+k}) \in B_2$
then $X(\zeta_{n-1+m}) \in A_1$ and $X(\zeta_{n+m}) \in A_2$ for some $0 \leq m < k$.

(We assume that for all $x \in A_2$ there exists $x \in A_1$ such that $x_1 \to x_2$ and that for all $x_2 \in B_2$ there exists $x_1 \in B_1$ such that $x_1 \to x_2$.)

Next we define two sequences of random times, $\{S_j : j \geq 0\}$ and $\{T_j : j \geq 1\}$, where S_{j-1} is the start time of the jth passage time for the marked job and T_j is the termination time of this jth passage time. Assuming that the initial state of the augmented job stack process X is such that a passage time for the marked job begins at $t = 0$, set $S_0 = 0$,

$$S_j = \inf \{\zeta_n \geq T_j : X(\zeta_n) \in A_2, X(\zeta_{n-1}) \in A_1\},$$

and

$$T_j = \inf \{\zeta_n > S_{j-1} : X(\zeta_n) \in B_2, X(\zeta_{n-1}) \in B_1\},$$

$j \geq 1$. Then the jth passage time for the marked job is $P_j = T_j - S_{j-1}$. (We assume that the process $\{X(S_n) : n \geq 0\}$ is aperiodic.)

The argument in Appendix 3 shows that the sequence of passage times for any other job (as well as the sequence of passage times, irrespective of job identity, in order of start or termination) converges in distribution to the same random variable as the sequence of passage times for the marked job.

Example 3.3. (Cyclic Queues) Suppose that the set, C, of (center, class) pairs is $C = \{(1,1),(2,1)\}$. Let $Z(t)$ be the number of jobs waiting or in service at center 1 at time t and set $X(t) = (Z(t),N(t))$, where $N(t)$ is the position of the marked job in the job stack at time t. Thus, for example, the augmented job stack process is in state $(0,N)$ if all N jobs are at center 2 and the marked job is in service. Upon completion of service at center 2, the marked job goes into service at center 1, the remaining $N-1$ jobs are at center 2, and the process is in state $(1,1)$. The state space, G, of the augmented job stack process $X = \{X(t): t \geq 0\}$ is

$$G = \{(z,n): 0 \leq z \leq N, 1 \leq n \leq N\}.$$

Consider the passage time that starts when a job completes service at center 2 (and joins the tail of the queue at center 1) and terminates when the job next joins the tail of the queue at center 2. This passage time (denoted by P) is specified by the sets

$$A_1 = \{(i,N): 0 \leq i < N\},$$

$$A_2 = \{(i,1): 0 < i \leq N\},$$

$$B_1 = \{(i,i): 0 < i \leq N\},$$

and

$$B_2 = \{(i-1,i): 0 < i \leq N\}.$$

Figure 4.1 shows state transitions for the process X and the sets A_1, A_2, B_1, and B_2 for $N = 2$ jobs. In the schematic representation of the job stack, jobs to the left of the vertical bar are at center 1 and jobs to the right are at center 2. The symbol × denotes the marked job and ○ denotes an unmarked job.

150 4 Non-Markovian Networks

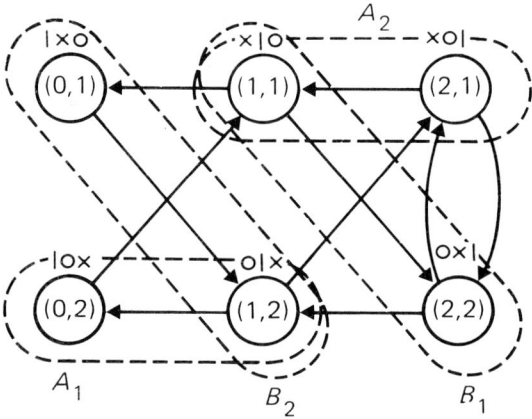

Figure 4.1. Subsets of G for passage time P

Recall that for $z \in D^*$, $U(z)$ is the set of all (center, class) pairs $(i,j) \in C$ such that in state z there is a job of class j in service at center i. Analogously, define $U(x)$ for $x \in G^*$: $U(x) = U(z)$ when $x = (z,n)$ for some $z \in D^*$ and $n \in \{1,2,...,N\}$. For $x,x' \in G^*$ and $u = (i,j) \in U(x)$, we denote by $p(x';x,u(m))$ the probability that the augmented job stack process X makes a transition to state x' given that in state x there is a completion of service to a job of class j by server m at center i. (We write $x \to x'$ if $p(x';x,u(m)) > 0$.) We say that x' is *accessible from* x and write $x \sim x'$ if there exists a finite sequence $x_1, x_2, ..., x_n$ of states of the augmented job stack process, (center, class) pairs $u_{i_0}, u_{i_1}, \ldots, u_{i_n}$, and servers $m_0, m_1, \ldots, m_n$ such that

$$p(x_1; x, u_{i_0}(m_0)) p(x_2; x_1, u_{i_1}(m_1)) \cdots p(x'; x_n, u_{i_n}(m_n)) > 0.$$

When $x \sim x'$ and $x' \sim x$ we say that *x and x' communicate*.

In the absence of some restriction on the building blocks of a network of queues with priorities among job classes, the sequence of passage times for the marked job need not converge in distribution to a random variable independent of the initial state of the system. We assume that for some state, z^*, of the job stack process the sets $D = \{z \in D^* : z^* \sim z\}$ and $G = \{(z,n) \in G^* : z \in D\}$ are irreducible in the sense that any pair of states communicate. For networks with more than one service center ($s > 1$), it is sufficient that for some service center, i_0, either $k(i_0) = 1$ or service at center i_0 to a job of class $j_{k(i_0)}(i_0)$ (the lowest priority job class seen by center i_0) is preempted when any other job of higher priority joins the queue. Let $z_{i_0}^*$ be the state of the job stack process in which there is one job of class $j_{k(i_0)}(i_0)$ in service at center i_0 and $N - 1$ jobs of class $j_{k(i_0)}(i_0)$ in queue at center i_0 (or in service if center i_0 is a multiple server center). Then take D to be the set of all states of the job stack process Z that are accessible from $z_{i_0}^*$:

(3.4) $$D = \{z \in D^* : z_{i_0}^* \sim z\})$$

and set

(3.5) $$G = \{(z,n) \in G^* : z \in D\}.$$

The argument used to establish Proposition 3.4 of Chapter 3 shows that $x_{i_0}^* = (z_{i_0}^*, N)$ can serve as a target state for the augmented job stack process in the sense that x and $x_{i_0}^*$ communicate for all $x \in G$.

Proposition 3.6. Suppose that the number of service centers $s > 1$. Also suppose that the routing matrix P is irreducible and that for some service center, i_0, either $k(i_0) = 1$ or service to a job of class $j_{k(i_0)}(i_0)$ at center i_0 is preempted when any other job of higher priority joins the queue. Then x and x' communicate for all $x, x' \in G$.

Henceforth, we assume that the subsets A_1, A_2, B_1, and B_2 that define the starts and terminations of passage passage times for the marked job are subsets of G. Without loss of generality, we also suppose that $X(0) \in G$; thus, we consider simulation of the augmented job stack process restricted to the set G.

Set

$$S = \{(x_1, x_2) : x_1 \in A_1, x_2 \in A_2 \text{ and } p(x_2; x_1, u(m)) > 0 \text{ for some } u(m)\}.$$

The entrances of the augmented job stack process X to the set S correspond to the starts of passage times for the marked job. Recall that we write $h(z, n) = (i, j)$ when the job in position n in the job stack associated with state $z \in D$ is of class j at center i. Now define a subset, S', of S according to

(3.7) $\quad S' = \{(z, N, z', n') \in S :$ for some single server center i

and some $(i, j_l(i)) \in C$, $h(z, N) = (i, j_l(i))$ and

$h(z, n) = (i, j_{l_n}(i))$ with $l_n \geq l$, $n = 1, 2, \ldots, N-1\}$

and assume that $S' \neq \emptyset$. According to this definition there exists a state, (z, N), of the augmented job stack process such that (i) the marked job is in service at some single server center, (ii) the other $N - 1$ jobs are in queue at the center as jobs of equal or lower priority, and (iii) with positive probability a passage time for the marked job starts upon completion of the service in progress. These "single states" are used to define regenerative cycles for the marked job method developed in the following sections.

Definition 3.8. An element $x = (z, N) \in A_1$ is called a *single state of the augmented job stack process for the passage time specified by the sets* A_1, A_2, B_1, and B_2 if for some single server center, i_0, and some (center, class) pair $(i_0, j_l(i_0)) \in C$

(i) $h(z,N) = (i_0, j_l(i_0))$;
(ii) $h(z,n) = (i_0, j_{l_n}(i_0))$ with $l_n \geq l$, $n = 1,2,...,N - 1$; and
(iii) there exists $(z',n') \in A_2$ such that $(z,N) \to (z',n')$.

(Recall that $j_1(i_0), j_2(i_0),...,j_{k(i_0)}(i_0)$ are the classes of jobs served at center i_0. Definition 3.8 requires that the class, $j_{l_n}(i_0)$, at center i_0 of the job in position n satisfy $j_{l_n}(i_0) \in \{j_l(i_0), j_{l+1}(i_0),...,j_{k(i_0)}(i_0)\}$ for all $n < N$.)

4.4 Recurrence and Regeneration

The GSMP associated with the augmented job stack process has state space, G, and event set, E, where event $e =$ "service completion to a job of class j by server m at center i" is active in state x ($e \in E(x)$) if and only if (center, class) pair $(i,j) \in U(x)$ and in state x a job of class j receives service from server m at center i. Jumps of the process are governed by the probability mass function $p(\cdot;x,e)$. At a jump from state x to state x' triggered by event e, new clock values are generated for each $e' \in E(x') - (E(x) - \{e\})$. The distribution of such a new clock value is denoted by $F(\cdot;x',e',x,e)$; by assumption, each has finite mean and a density function that is continuous and positive on $(0, +\infty)$. For $e' \in E(x') \cap (E(x) - \{e\})$, the old clock reading is kept after the jump. For $e' \in (E(x) - \{e\}) - E(x')$, event e' ceases to be scheduled after the jump. Denote the set of all possible events that can occur by $E = \{e_1, e_2,...,e_M\}$. With each $x \in G$, associate the set of clock readings

$$C(x) = \{(c_1, c_2,...,c_M): c_i \geq 0 \text{ and } c_i > 0 \text{ if and only if } e_i \in E(x)\},$$

where c_i is the reading on the clock corresponding to event e_i. Also observe that the GSMP has unit speeds and is irreducible in the sense that x and x' communicate for all $x, x' \in G$.

Proposition 4.1. Suppose that the number of service centers $s > 1$, the routing matrix P is irreducible, and for some service center, i_0, either $k(i_0) = 1$ or service to a job of class $j_{k(i_0)}(i_0)$ at center i_0 is preempted when any other job of higher priority joins the queue. Also suppose that $S' \neq \emptyset$ and set

(4.2) $\quad A_2' = \{(z',n') \in A_2: (z,N,z',n') \in S' \text{ for some } z \in D\}.$

Then $P\{X(S_n) = x' \text{ i.o.}\} = 1$ for any $x' \in A_2'$.

Proof. By Proposition 2.1 it is sufficient to show that (i) the GSMP is irreducible and (ii) each clock setting distribution has finite mean and a density function that is continuous and positive on $(0, +\infty)$. Condition (i) holds as a consequence of Proposition 3.6 and condition (ii) holds by assumption. $\square$

Select an element x' from the set A_2' of (4.2). By Proposition 2.1 we know that $\{X(S_n): n \geq 0\}$ makes a transition to state x' infinitely often with probability one. Furthermore, at such a start time, S_n, the only (two) clocks that are active were just set, as described in the proof of Proposition 2.4. Since the jumps of the augmented job stack process X are governed by the Markovian mass function $p(\cdot; x, e)$ and the only active clocks have just been set at time S_n, the future evolution of the process X is independent of the history of the process before time S_n and has the same distribution as it does when $X(0) = x'$. Thus, the subsequence of start times S_n at which $X(S_n) = x'$ are regeneration points for the process X. Since the state space, G, of the process is finite and the clock setting distributions have finite mean, the expected time between regeneration points is finite.

Proposition 4.3 The augmented job stack process $X = \{X(t): t \geq 0\}$ is a regenerative process in continuous time and the expected time between regeneration points is finite.

By the argument leading to Proposition 4.3, it is clear that the random indices β_n such that $X(S_{\beta_n}) = x'$ form a sequence of regeneration points for the process $\{(X(S_n), P_{n+1}): n \geq 0\}$; this follows from the fact that passage times P_{n+1} start from scratch when $X(S_{\beta_n}) = x'$.

Proposition 4.4. The process $\{(X(S_n), P_{n+1}): n \geq 0\}$ is a regenerative process in discrete time and the expected time between regeneration points is finite.

Since $\{X(S_n): n \geq 0\}$ is aperiodic, Proposition 4.4 implies that

$$(X(S_n), P_{n+1}) \Rightarrow (X, P)$$

as $n \to \infty$. This means that there exist random variables X and P such that

$$\lim_{n \to \infty} P\{X(S_n) = x, P_{n+1} \leq y\} = P\{X = x, P \leq y\} \equiv F(i,y)$$

for all $x \in A_2$ and $y \in [0, +\infty)$ for which $F(x,\cdot)$ is continuous. The goal of the simulation is the estimation of $r(f) = E\{f(P)\}$, where f is a real-valued (measurable) function.

4.5 The Marked Job Method

To obtain estimates for $r(f)$, select $x' \in A_2'$ and begin a simulation of the augmented job stack process X with $X(0) = x'$. Carry out the simulation of the process in cycles defined by the successive times at which the process $\{X(S_n): n \geq 0\}$ makes a transition to state x'. Denote by α_m the length (in discrete time units) of the mth cycle of $\{X(S_n): n \geq 0\}$; α_m is the number of passage times for the marked job in the mth cycle. Set $\beta_0 = 0$ and $\beta_m = \alpha_1 + \ldots + \alpha_m$, $m \geq 1$. Also set

$$Y_m(f) = \sum_{j=\beta_{m-1}+1}^{\beta_m} f(P_j).$$

Propositions 5.1 and 5.2 follow from Proposition 4.4 and general results for regenerative processes.

Proposition 5.1. The sequence of pairs of random variables $\{(Y_m(f), \alpha_m) : m \geq 1\}$ are independent and identically distributed.

Proposition 5.2. Provided that $P\{P \in D(f)\} = 0$ and $E\{|f(P)|\} < \infty$,

$$r(f) = \frac{E\{Y_1(f)\}}{E\{\alpha_1\}}.$$

With these results

(5.3) $$\hat{r}(n) = \frac{\bar{Y}(n)}{\bar{\alpha}(n)} = \frac{\sum_{m=1}^{n} Y_m(f)}{\sum_{m=1}^{n} \alpha_m}$$

is a strongly consistent point estimate for $r(f)$ and an asymptotic $100(1 - 2\gamma)\%$ confidence interval is

(5.4) $$\left[\hat{r}(n) - \frac{z_{1-\gamma} \, s(n)}{\bar{\alpha}(n) \, n^{1/2}}, \, \hat{r}(n) + \frac{z_{1-\gamma} \, s(n)}{\bar{\alpha}(n) \, n^{1/2}}\right].$$

The quantity $s^2(n)$ is a strongly consistent point estimate for $\sigma^2(f) = \mathrm{var}\,(Y_1(f) - r(f)\alpha_1)$. Asymptotic confidence intervals for $r(f)$ are based on the c.l.t.

$$\frac{n^{1/2} \{\hat{r}(n) - r(f)\}}{\sigma(f)/E\{\alpha_1\}} \Rightarrow N(0,1)$$

as $n \to \infty$.

Algorithm 5.5. (Marked Job Method for Non-Markovian Networks)

1. Select a single state, $x = (z, N)$, of the augmented job stack process and a state $x' = (z', n') \in A_2'$ so that a passage time for the marked job starts when the process makes a transition from state x to state x'.

2. Set $X(0) = x'$ and simulate the augmented job stack process. Observe a fixed number, n, of cycles defined by the successive times at which $X(S_n) = x'$. In each cycle measure all passage times for the marked job.
3. Compute the number, α_m, of passage times for the marked job in the mth cycle and the quantity

$$Y_m(f) = \sum_{j=\beta_{m-1}+1}^{\beta_m} f(P_j),$$

where $\beta_0 = 0$ and $\beta_m = \alpha_1 + ... + \alpha_m$.
4. Form the point estimate

$$\hat{r}(n) = \frac{\overline{Y}(n)}{\overline{\alpha}(n)}.$$

5. Form the asymptotic $100(1 - 2\gamma)\%$ confidence interval

$$\left[\hat{r}(n) - \frac{z_{1-\gamma} \, s(n)}{\overline{\alpha}(n) \, n^{1/2}}, \, \hat{r}(n) + \frac{z_{1-\gamma} \, s(n)}{\overline{\alpha}(n) \, n^{1/2}} \right].$$

The marked job method for networks with general service times requires selection of a return state $x' \in A'_2 \subseteq G$. As $G = \{(z,n) \in G^* : z \in D\}$, it is sufficient to determine the elements of the set D. It is easy to show that the elements of D are precisely those states of the job stack process that are accessible from some state z_i^*. For $i = 1,2,...,s$, let z_i^* be the state of the job stack process in which there is one job of class $j_{k(i)}(i)$ in service at center i and $N - 1$ jobs of class $j_{k(i)}(i)$ in queue at center i (or in service if center i is a multiple server center).

Proposition 5.6. Assume that the routing matrix P is irreducible and that for some service center, i_0, either $k(i_0) = 1$ or service to a job of class $j_{k(i_0)}(i_0)$ at center i_0 is preempted when any other job of higher priority joins the queue. Let $z \in D^*$. Then $z \in D$ if and only if $z_i^* \sim z$ for some $i = 1,2,\ldots,s$.

Proof. Without loss of generality, assume that $i_0 = 1$: either $k(1) = 1$ or service to a job of class $j_{k(1)}(1)$ at center 1 is preempted when any other job of higher priority joins the queue. First observe that $z_i^* \in D$, $i = 1,2,\ldots,s$. (By the argument in the proof of Proposition 3.6, $z \sim z_1^*$ for all $z \in D^*$. This implies that $z_1^* \in D$. By hypothesis, either $k(1) = 1$ or service to jobs of class $j_{k(1)}(1)$ at center 1 is preempted when any other job of higher priority joins the queue. In either case, it is easy to show that z_i^* is accessible from z_1^*, $i = 1,2,\ldots,s$. This implies that $z_i^* \in D$.) Therefore, $z \in D$ if $z_i^* \sim z$ for some i. Conversely, pairs of states $z, z' \in D$ communicate by Proposition 3.6. Therefore $z_i^* \sim z$ for all i if $z \in D$. □

Example 5.7. (Cyclic Queues With Feedback) Let $Z(t)$ be the number of jobs waiting or in service at center 1 at time t. For fixed $0 < p < 1$ the routing matrix P is

$$P = \begin{matrix} p & 1-p \\ 1 & 0 \end{matrix}.$$

Then $D = D^*$ and $G = G^*$. Consider (i) the passage time, R, that starts when a job enters the center 1 queue upon completion of service at center 2 and terminates the next such time at which the job joins the center 1 queue and (ii) the passage time, P, that starts when a job enters the center 1 queue upon completion of service at center 2 and terminates when the job next joins the center 2 queue. For these passage times,

$$A_1 = \{(i,N): 0 \le i < N\}$$

and

$$A_2 = \{(i,1): 0 < i \leq N\}.$$

The set S that defines the starts of passage times for the marked job is

$$S = \{(i,N,i+1,1): 0 \leq i < N\}.$$

The subset $S' = \{(0,N,1,1)\}$ and the set $A_2' = \{(1,1)\}$.

Example 5.8. (Cyclic Queues With Nonpreemptive Priority) Consider a network with two service centers and two job classes. Assume that the set, C, of (center, class) pairs is $C = \{(1,1),(2,1),(2,2)\}$ and jobs of class 2 have nonpreemptive priority over jobs of class 1 at center 5. Set

(5.9) $$Z(t) = \left(Q_1(t), C_1^{(2)}(t), C_2^{(2)}(t), S_2(t)\right).$$

Suppose that the irreducible routing matrix P is

$$P = \begin{matrix} 0 & 1 & 0 \\ 0 & 0 & 1 \\ 1 & 0 & 0 \end{matrix}.$$

Also suppose that for the passage time of interest

$$A_1 = \{(q,c_1,c_2,s,N) \in G : s = 2\}$$

and

$$A_2 = \{(q,c_1,c_2,s,1) \in G : q > 0\}.$$

Thus, a passage time starts when a job of class 2 completes service at center 2 and joins the center 1 queue (as class 1). Note that the set $D^* - D \neq \emptyset$. For this network, $z_2^* = (0, N-1, 0, 1)$ and $U(z_2^*) = \{(2,1)\}$. By Proposition 5.6, state $z = (0, N-1, 0, 2) \in D$ since $q(z; z_2^*, u(1)) = 1$ with $u = (2,1)$. It follows that state

$(0,N-1,0,2,N,1,N-2,0,1,1) \in S'$ and can be used to define cycles for the passage time simulation.

Tables 4.1 and 4.2 give point and interval estimates obtained using the marked job method for the expected value and percentiles of the passage time R, in the data base management system model. This passage time is specified by four subsets $(A_1, A_2, B_1,$ and $B_2)$ of the set G:

$$A_1 = \{N - (i + 1),0,0,0,0,0,7,i,N - i):0 \leq i < N\}$$

and

$$A_2 = \{N - (i + 1),0,0,0,0,0,6,i,N - i):0 \leq i < N\}$$

with $B_1 = A_1$ and $B_2 = A_2$. The passage time starts when a job completes service at center 1 as class 7 and terminates when the job next completes service at center 1 as class 7. Set $z_0 = (N - 1,0,0,0,0,0,7,0)$ and $z_0' = (N - 1,0,0,0,0,0,6,0)$. The subset S' is

$$S' = \{(z_0, N, z_0', N)\}$$

so that there is one single state, (z_0, N), for the passage time. When the augmented job stack process is in state (z_0, N) all N jobs are of class 7 at center 1 and the marked job is in service.

All service times are exponentially distributed. The mean, λ_j^{-1}, of the service time distribution depends on the class, j, of the job in service. Exponential service times have been generated by logarithmic transformation of uniform random numbers obtained as in (4.12) of Chapter 3. Independent streams of exponential random numbers (resulting from different seeds of the uniform random number generator) were used to generate exponential service time sequences for the individual job classes.

4.5 The Marked Job Method

Table 4.1 Simulation Results for Passage Time R in Data Base Management System Model

$$N = 2, p_1 = 0.1, p_2 = 0.2, \lambda_1^{-1} = 50.0,$$
$$\lambda_2^{-1} = \lambda_4^{-1} = 3.3, \lambda_3^{-1} = \lambda_5^{-1} = 1.5, \lambda_6^{-1} = 6.7, \lambda_7^{-1} = 1.0$$

Exponential Service Times for All Job Classes
Return State is (1,0,0,0,0,0,6,0,2)

No. of cycles, n	200	400	600	800	1000
Simulated time	33689.38	80632.23	129362.60	174804.60	219304.30
No. of transitions/cycle	52.18	59.98	63.35	64.33	64.47
Fraction of time center 1 busy	0.7770 ±0.0524	0.7528 ±0.0289	0.7433 ±0.0219	0.7480 ±0.0180	0.7484 ±0.0157
Fraction of time center 2 busy	0.5282 ±0.0603	0.5747 ±0.0321	0.5894 ±0.0241	0.5837 ±0.0199	0.5839 ±0.0174
$\bar{a}(n)$	5.135	5.400	5.513	5.541	5.589
$E\{R\}$ (84.556)	78.8978 ±8.0073	83.9919 ±5.3648	85.7842 ±4.3583	85.9836 ±3.5606	84.7062 ±3.0340
$P\{R \leq 24.14\}$	0.1897 ±0.0324	0.2010 ±0.0226	0.2042 ±0.0189	0.1899 ±0.0157	0.1943 ±0.0139
$P\{R \leq 45.28\}$	0.4145 ±0.0416	0.4177 ±0.0281	0.4178 ±0.0217	0.4097 ±0.0188	0.4125 ±0.0165
$P\{R \leq 84.56\}$	0.7424 ±0.0337	0.6990 ±0.0246	0.6916 ±0.0198	0.6837 ±0.0170	0.6840 ±0.0149
$P\{R \leq 169.14\}$	0.8829 ±0.0258	0.8615 ±0.0182	0.8561 ±0.0148	0.8608 ±0.0124	0.8617 ±0.0111
$P\{R \leq 338.25\}$	0.9742 ±0.0142	0.9729 ±0.0092	0.9675 ±0.0081	0.9685 ±0.0067	0.9718 ±0.0056

Under the exponential service time assumptions the theoretical value (given in parentheses in Table 4.1) for $E\{R\}$ can be obtained. The initial state for the augmented job stack process X (and return state identifying regenerative cycles) is $x' = (N - 1,0,0,0,0,0,6,0,N)$. For $N = 2$ jobs and 200 cycles, the resulting point estimate for $E\{R\}$ is 78.8978 and the half length of the 90% confidence interval is 8.0076. For 200, 400, 600, 800 and 1000 cycles, the confidence intervals contain the theoretical value. Estimates for $N = 4$ jobs are given in Table 4.2.

Table 4.2 Simulation Results for Passage Time R in Data Base Management System Model

$N = 4$, $p_1 = 0.1$, $p_2 = 0.2$, $\lambda_1^{-1} = 50$,

$\lambda_2^{-1} = \lambda_4^{-1} = 3.3$, $\lambda_3^{-1} = \lambda_5^{-1} = 1.5$, $\lambda_6^{-1} = 6.7$, $\lambda_7^{-1} = 1.0$

Exponential Service Times for All Job Classes
Return State is (3,0,0,0,0,0,6,0,4)

No. of cycles, n	200	400	600	800	1000
Simulated time	145268.80	270127.30	406800.50	537384.00	683064.00
No. of transitions/cycle	249.06	231.85	233.82	232.62	236.36
Fraction of time center 1 busy	0.8755 ±0.0185	0.8758 ±0.0141	0.8801 ±0.0111	0.8849 ±0.0092	0.8817 ±0.0084
Fraction of time center 2 busy	0.7263 ±0.0266	0.7141 ±0.0197	0.7129 ±0.0156	0.7122 ±0.0132	0.7183 ±0.0116
$\bar{\alpha}(n)$	4.930	4.658	4.745	4.702	4.754
$E\{R\}$	147.3315 ±7.3648	144.9959 ±5.6797	142.8874 ±4.3073	142.8453 ±3.6739	143.6820 ±3.2584
$P\{R \leq 24.14\}$	±0.0923 ±0.0201	±0.0886 ±0.0139	±0.0861 ±0.0109	±0.0829 ±0.0095	±0.0852 ±0.0082
$P\{R \leq 45.28\}$	0.2039 ±0.0271	0.1997 ±0.0201	0.1939 ±0.0157	0.1877 ±0.0140	0.1906 ±0.0119
$P\{R \leq 84.56\}$	0.4320 ±0.0272	0.4283 ±0.0224	0.4254 ±0.0174	0.4242 ±0.0152	0.4222 ±0.0131
$P\{R \leq 169.14\}$	0.6907 ±0.0252	0.7112 ±0.0180	0.7201 ±0.0143	0.7188 ±0.0124	0.7137 ±0.0112
$P\{R \leq 338.25\}$	0.9077 ±0.0164	0.9152 ±0.0111	0.9203 ±0.0085	0.9203 ±0.0072	0.9178 ±0.0066

Comparison with Table 4.1 provides an indication of the effect on computational and statistical efficiency of the increase in the number of jobs. For simulations of equal length, the accuracy of the estimates for $E\{R\}$ is roughly comparable.

Table 4.3 Simulation Results for Passage Time R in Data Base Management System Model

$$N = 2, p_1 = 0.1, p_2 = 0.2, \lambda_1^{-1} = 50.0,$$

$$\lambda_2^{-1} = \lambda_4^{-1} = 1.3, \lambda_3^{-1} = \lambda_5^{-1} = 1.5, \lambda_6^{-1} = 6.7, \lambda_7^{-1} = 1.0$$

Exponential Service Times for Class 7 Jobs
Constant Service Times for Other Job Classes
Return State is (1,0,0,0,0,0,6,0,2)

No. of cycles, n	200	400	600	800	1000
Simulated time	45291.23	92218.60	143658.30	193607.10	241766.60
No. of transitions/cycle	68.81	67.39	69.75	70.28	70.22
Fraction of time center 1 busy	0.7727 ±0.0256	0.7425 ±0.0205	0.7386 ±0.0171	0.7360 ±0.0149	0.7360 ±0.0135
Fraction of time center 2 busy	0.6071 ±0.0312	0.6193 ±0.0229	0.6190 ±0.0196	0.6198 ±0.0169	0.6195 ±0.0152
$\bar{a}(n)$	5.795	5.762	5.765	5.751	5.755
$E\{R\}$	81.0219 ±6.4320	83.4557 ±4.4219	86.5933 ±3.6403	87.9632 ±3.2392	87.7556 ±2.8777
$P\{R \leq 24.14\}$	0.1843 ±0.0299	0.1828 ±0.0206	0.1706 ±0.0170	0.1699 ±0.0147	0.1699 ±0.0130
$P\{R \leq 45.28\}$	0.4741 ±0.0370	0.4588 ±0.0250	0.4394 ±0.0203	0.4321 ±0.0181	0.4276 ±0.0161
$P\{R \leq 84.56\}$	0.7048 ±0.0303	0.6914 ±0.0217	0.6745 ±0.0185	0.6674 ±0.0162	0.6708 ±0.0144
$P\{R \leq 169.14\}$	0.8712 ±0.0228	0.8606 ±0.0157	0.8517 ±0.0134	0.8487 ±0.0117	0.8505 ±0.0103
$P\{R \leq 338.25\}$	0.9696 ±0.0122	0.9692 ±0.0086	0.9668 ±0.0070	0.9668 ±0.0060	0.9673 ±0.0053

Table 4.3 gives point estimates and 90% confidence intervals for the expected value and percentiles of the passage time R when service times to jobs of class 7 are exponentially distributed and service times to the other job classes are constant. Parameter values are as in Table 4.2. Although the positivity hypothesis used in the proof of Proposition 4.1 is not satisfied, it can be shown for this network that the process $P\{X(S_n) = x'\text{ i.o.}\} = 1$ and that the marked job method is valid.

4.6 Finite Capacity Open Networks

We now consider open networks of service centers. Jobs arrive at the network, traverse the network and receive service along the way, and finally depart from the network. The network structure we permit is essentially the same as that described in Section 4.1, except that here the networks are open; thus, there are arrivals from an external source and departures to an external sink. Only a finite number of jobs, N, may reside in the network at a given time.

We consider two formulations of finite capacity open networks. In *arrival process shutdown* models, a job arriving when the network already contains $N - 1$ jobs causes the arrival process to shut down; the arrival process remains shut down until the first subsequent departure from the network. In *jobs turned away* models, the arrival process never shuts down, but jobs arriving when the network already contains N jobs are lost.

The arrival processes we allow are particular stochastic point processes (series of events) associated with time-homogeneous Markov renewal processes (MRP's). Let J be a finite or countable set, $W = \{W_n : n \geq 0\}$ be random variables assuming values in J, and $U = \{U_n : n \geq 0\}$ be nonnegative random variables such that $0 = U_0 \leq U_1 \leq U_2 \ldots$. Recall that a stochastic process $(W, U) = \{(W_n, U_n) : n \geq 0\}$ is a MRP provided that

$$P\{W_{n+1} = j, U_{n+1} - U_n \le t \mid W_0,\ldots,W_n; U_0,\ldots,U_n\}$$

$$= P\{W_{n+1} = j, U_{n+1} - U_n \le t \mid W_n\}$$

with probability one for all $n \ge 0$, $j \in J$, and $t \ge 0$, and is time-homogeneous provided that $P\{W_{n+1} = j, U_{n+1} - U_n \le t \mid W_n\}$ is independent of n. Let $M = \{M(t): t \ge 0\}$ be the SMP generated by the process (W,U):

(6.1) $$M(t) = W_n \quad \text{for } U_n \le t < U_{n+1}.$$

By definition, a *Markov arrival process* is a stochastic point process U that satisfies the condition

(6.2) $$P\{U_{n+1} - U_n \le t \mid W_n = i\} = F_i(t)$$

for all $i \in J$ and $t \ge 0$, where F_i is the distribution of a positive random variable with finite mean. Throughout, we assume that M is irreducible and positive recurrent and that the MRP (W,U) is independent of the service times and Markovian routing of jobs in the network. Note that in arrival process shutdown models we can think of the arrival process as operating in virtual time: the nth job arrives at virtual time U_n. The actual time of the nth arrival, however, may be later due to the finite capacity constraint. In jobs turned away models, the nth job arrives at time U_n but may be turned away.

Example 6.3. (Renewal Process) Let $J = \{1\}$ so that $W_n = 1$ with probability one for all $n \ge 0$, and let

$$P\{U_{n+1} - U_n \le t\} = 1 - e^{-\lambda t}$$

for all $t \ge 0$, where $\lambda > 0$. Then the Markov arrival process $U = \{U_n: n \ge 0\}$ is a renewal process.

Example 6.4. (Switching Poisson Process) Let $J = \{1,2,...,k\}$ and

$$q_{ij} = P\{W_{n+1} = j \mid W_n = i\}$$

for $i,j \in J$. Also let

$$P\{U_{n+1} - U_n \leq t \mid W_n = i\} = 1 - e^{-\lambda_i t}$$

for all $t \geq 0$ and $i \in J$, where $\lambda_i > 0$. The successive times-between-events in the Markov arrival process $U = \{U_n : n \geq 0\}$ are exponentially distributed with parameters governed by the transition matrix $\{q_{ij} : i,j \in J\}$.

Example 6.5. (Branching Poisson Process) The branching Poisson process is a model for clustered arrivals and is constructed as follows. A Poisson process with parameter λ_1 generates a series of primary events, and with independent probability $r \in (0,1]$, a primary event initiates a subsidiary series of events. Each subsidiary process consists of a geometrically distributed (mean $(1-p)^{-1}$) number of subsidiary events, and the times between these subsidiary events are independent and exponentially distributed with parameter λ_2. Finally, the branching Poisson process is the superposition of all primary and subsidiary events.

To represent this process as a Markov arrival process, we set $J = \{0,1,2,...\}$ and identify W_n with the number of subsidiary processes active at the time of the nth event of the branching Poisson process. Then for $i \in J$ and $t \geq 0$,

$$P\{W_{n+1} = j \mid W_n = i\} = \begin{cases} i\lambda_2(1-p)/(\lambda_1 + i\lambda_2), & j = i-1 \\ [i\lambda_2 p + (1-r)\lambda_1]/(\lambda_1 + i\lambda_2), & j = i \\ r\lambda_1/(\lambda_1 + i\lambda_2), & j = i+1 \end{cases}$$

and

$$P\{U_{n+1} - U_n \leq t \mid W_n = i\} = 1 - e^{-(\lambda_1 + i\lambda_2)t}.$$

We permit a finite number of single-server service centers, s, and a finite number of (mutually exclusive) job classes, c. At every time epoch each job is in exactly one job class, but jobs may change class as they traverse the network. Upon completion of service at center i, a job of class j goes to center k and changes to class l with probability $p_{ij,kl}$,

$$P = \{p_{ij,kl}: (i,j),(k,l) \in C\}$$

is a given irreducible stochastic matrix and $C \subseteq \{1,2,...,s\} \times \{1,2,...,c\}$ is the set of (center, class) pairs in the network. For $(i,j) \in C$, a job from the external source arrives at center i as a job of class j with probability p_{ij}; a job of class j completing service at center i departs to the external sink with probability

$$q_{ij} = 1 - \sum_{(k,l) \in C} p_{ij,kl}.$$

At each service center jobs queue and receive service according to a fixed priority scheme among classes; the priority scheme may differ from center to center. Within a class at a center, jobs receive service according to a fixed discipline, and, some centers may never see jobs of certain classes as determined by the routing matrix, P. A job in service at a center may or may not be preempted if another job of higher priority joins the queue at the center. All service times in the network are mutually independent, and at a particular center have an exponential distribution with parameter which may depend on the service center, class of job being served, and the state of the job stack process.

As before, we let $S_i(t)$ be the class of job receiving service at center i at time t, $i = 1,2,...,s$; $S_i(t) = 0$ if at time t there are no jobs at center i. The classes of jobs served at center i ordered by decreasing priority are $j_1(i), j_2(i),...,j_{k(i)}(i)$, all elements of the set $\{1,2,..,c\}$. Let $C^{(i)}_{j_1}(t), C^{(i)}_{j_2}(t),...,C^{(i)}_{j_{k(i)}}(t)$ be the number of jobs in queue at time t of the various classes of jobs served at center i. We order the (at most N) jobs in the network in a linear stack and define the state of the system at time t to be the vector $Z(t)$ given by

$$Z(t) = \bigl(C^{(1)}_{j_{k(1)}}(t),\ldots,C^{(1)}_{j_1}(t),S_1(t);\ldots;C^{(s)}_{j_{k(s)}}(t),\ldots,C^{(s)}_{j_1}(t),S_s(t)\bigr).$$

The job stack at time t corresponds to the nonzero components in the vector $Z(t)$ and is an ordering of the jobs by class at the individual centers. Within a class at a particular service center, jobs waiting appear in order of their arrival at the center, the latest to arrive being closest to the top of the stack.

To deal with passage times, we again introduce a marked job and denote by $N(t)$ the position (from the top) of the marked job in the job stack at time t. Recalling that $M = \{M(t): t \geq 0\}$ is the SMP associated with the Markov arrival process, set

(6.6) $$X(t) = (M(t), Z(t), N(t))$$

and call $X = \{X(t): t \geq 0\}$ the *augmented job stack process*.

The principal concern here remains the estimation of general characteristics of passage times, the times required for a job to traverse a specified portion of the open network. To estimate passage times, we track an appropriate sequence of typical jobs, based on the idea of a marked job, and measure the passage times for a sequence of marked jobs. These are to be typical jobs in the sense that the sequence of passage times for the marked jobs should converge in distribution to the same random variable as do the passage times for all jobs. It is necessary to take some care to ensure that this is the case.

Our *job marking scheme* is as follows. Let m be the number of jobs left behind by the marked job at the epoch at which it departs from the network (i.e., goes to the external sink), $m = 0, 1, \ldots, N - 1$. Then for arrival process shutdown models, the $(N - m)$th subsequent arrival is the next marked job. For jobs turned away models, the $(N + 1 - m)$th subsequent arrival is the next marked job.

Note that this marking scheme ensures that there is at most one marked job in the network at a time, and thus there is no need for further augmentation of the numbers-in-queue state vector for the measurement of passage times. (If there is no marked job in the network at time t, we define $N(t)$ to be zero.) Note also that more than one passage time can start (and terminate) for a particular marked job before it departs from the network.

We claim that the sequence of passage times for jobs marked by this scheme has the desired property. For arrival process shutdown models, introduce a so-called *phantom server* that generates the times $U = \{U_n : n \geq 0\}$ according to the MRP (W, U) of the Markov arrival process. Denoting by m ($0 \leq m < N$) the number of jobs left behind by the marked job at the instant it departs from the network, we route the marked job to the queue at the phantom server where $N - m - 1$ jobs are already present. Upon completion of service to a job by the phantom server, the job returns to the network in the same manner as arriving jobs: with probability p_{kl} the job goes to center k and becomes class l. See Figure 4.2.

This method generates arrivals to the network in exactly the same way the original arrival process does with the finite capacity constraint, namely, arrivals occur at the times U_n provided the number in the network is less than N. It is now clear that in effect we have a closed network of queues in which the marked job never leaves the network. Furthermore, the stochastic structure of the original problem remains. The chief advantage of this device is that we can use the marked job method of Section 3.6 for closed networks of queues. Note that there is no need for a "$\rho < 1$" condition to guarantee stability of the open network of queues since it is effectively a closed network with a finite number of jobs.

For jobs turned away models, we introduce a phantom server in a similar way. A job completing service at the phantom server with $m - 1$ other jobs at the server returns to the network with probability $1 - p(m)$, where $p(m) = 1$ if $m = 1$ and equals 0

170 4 Non-Markovian Networks

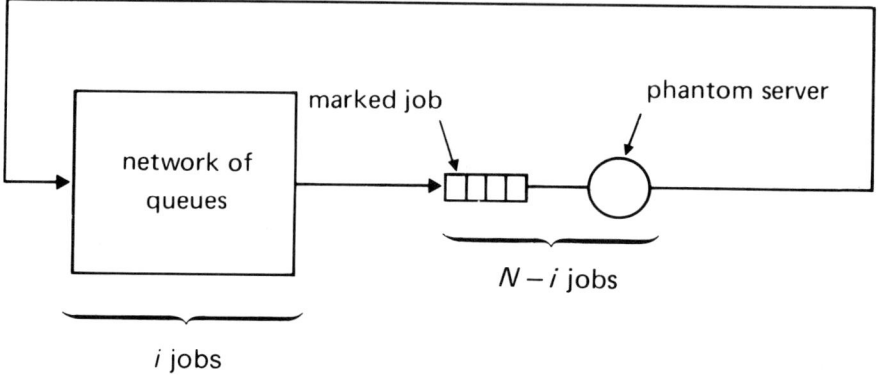

Figure 4.2. Flow of jobs with phantom server added.
"Arrival process shutdown" model

otherwise. The job joins the tail of the queue at the phantom server with probability $p(m)$. See Figure 4.3. In effect we are considering a closed network of queues with $N + 1$ jobs. It is straightforward to check that the marked job method of Section 4.5 also applies to closed networks of queues in which, as here, routing probabilities may depend on the number of jobs in the service center.

Having reduced the problem to estimation in a closed network of queues, it is necessary to modify the augmented job stack process X. View the phantom server as service center $s + 1$ (serving only one job class) and let $Q_{s+1}(t)$ be the total number of jobs at center $s + 1$ at time t. Then we augment the vector $Z(t)$ with the component $Q_{s+1}(t)$, and use this augmented $Z(t)$ to define $X(t)$ as in (6.6). We also modify the routing matrix P to describe

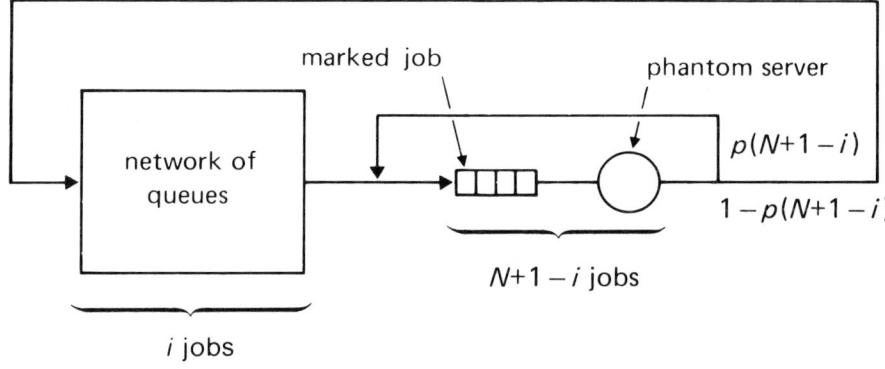

Figure 4.3. Flow of jobs with phantom server added. "Jobs turned away model"

the equivalent closed network of queues and assume that the resulting routing matrix is irreducible. The marked job method of Section 4.5 can be used to estimate passage times provided that the augmented job stack process X restricted to the set G is irreducible.

Example 6.7. (Tandem Queues With Feedback) Consider a finite capacity open network of queues with two service centers; see Figure 4.4. Jobs arrive from an external source and join the tail of the queue at center 4. Upon completion of service at center 1, a job joins the tail of the queue in center 1 (with probability p) or joins the tail of the queue in center 2 (with probability $1 - p$). Upon completion of service at center 2, a job goes to the external sink. Neither center 1 nor center 2 service is subject to interruption and both queue service disciplines are FCFS. Denote by R (resp., P) the (limiting) passage time that starts when a job arrives at the center 1 queue and terminates when the job departs from the system (resp., when the job enters the center 2 queue.)

Taking the set C of (center, class) pairs to be $C = \{(1,1),(2,2)\}$, the routing matrix P is

$$P = \begin{matrix} p & 1-p \\ 0 & 0 \end{matrix}.$$

The probabilities p_{ij} of jobs from the external source are $p_{11} = 1$ and $p_{22} = 0$; the probabilities q_{ij} of departure to the external sink are $q_{11} = 0$ and $q_{22} = 4$. The routing matrix for the equivalent closed network of queues is

$$\begin{matrix} p & 1-p & 0 \\ 0 & 0 & 1 \\ 1 & 0 & 0 \end{matrix}.$$

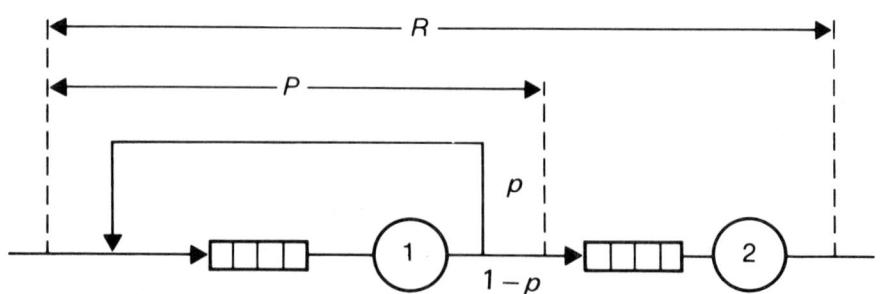

Figure 4.4. Tandem queues with feedback

As each center sees only one job class, set

$$Z(t) = (Q_1(t), Q_2(t)),$$

where $Q_i(t)$ is the number of jobs waiting or in service at center i at time t, $i = 1,2$. Let $N(t)$ be the position of the marked job in the job stack at time t and let $\{M(t): t \geq 0\}$ be the SMP associated with the Markov arrival process. Then set

$$X(t) = (M(t), Q_1(t), Q_2(t), N(t)).$$

The state space, G, of the augmented job stack process $X = \{X(t): t \geq 0\}$ is

$$G = \{(m, q_1, q_2, n): 0 \leq q_1, q_2, n \leq N; q_1 + q_2 \leq N; m \in J\},$$

where N is the maximum number of jobs in the network. The sets A_1 and A_2 that define the starts of the passage time, R, are

$$A_1 = \{(m, q_1, q_2, n) \in G : n = N\}$$

and

$$A_2 = \{(m, q_1, q_2, 1) \in G : q_1 > 0\}.$$

The sets B_1 and B_2 that define the terminations of the passage time are

$$B_1 = \{(m, q_1, q_2, n) \in G : n = q_1 + q_2, q_2 > 0\}$$

and

$$B_2 = \{(m, q_1, q_2, n) \in G : n = q_1 + q_2 + 1\}.$$

4.7 Passage Through Subnetworks

Passage times for all the jobs in a network can be recorded by observing a "fully augmented job stack process" that maintains the position of each of the jobs in the job stack. We show, under a mild restriction on the priorities among job classes, that the job stack process observed at the epochs at which passage times terminate is a regenerative process in discrete time. As a consequence, point and interval estimates for characteristics of limiting passage times can be obtained by observing a finite portion of a single sample path of the job stack process.

Terminations of passage times with no other passage times underway and exactly one job in service are regeneration points for the job stack process observed at termination times. In order for such epochs to exist we must exclude passage times that are complete circuits as well as passage times that always terminate with two or more jobs in service. A mild restriction on the priorities among job classes assures that infinitely many such epochs occur.

To obtain recurrence results, it is necessary to restrict the job stack process to a suitably chosen subset of its state space. We assume throughout that there exists $z^* \in D^*$ such that $z \sim z^*$ for all $z \in D^*$. This ensures that the set $D = \{z \in D^* : z^* \sim z\}$ of all states that are accessible from state z^* is irreducible in the sense that states z and z' communicate for all $z, z' \in D$. It is sufficient that for some service center, i_0, either $k(i_0) = 1$ or service at center i_0 to a job of class $j_{k(i_0)}(i_0)$ (the lowest priority job class seen by center i_0) is preempted when a job of higher priority joins the queue. Let $z^*_{i_0}$ be the state of the job stack process in which there is one job of class $j_{k(i_0)}(i_0)$ in service at center i_0 and $N - 1$ jobs of class $j_{k(i_0)}(i_0)$ in queue at center i_0. Then the set $D = \{z \in D^* : z^*_{i_0} \sim z\}$ of all states of the job stack process that are accessible from state $z^*_{i_0}$ is irreducible. We also assume that the sets A_1, A_2, B_1, and B_2 that define the starts and terminations of passage times are subsets of $G = \{(z,n) \in G^* : z \in D\}$.

Special single states of the job stack process are used explicitly in the labelled jobs method. Formally, define two sets S

and T according to

(7.1) $\qquad S = \{(x_1,x_2) : x_1 \in A_1,\ x_2 \in A_2,\ \text{and}\ x_1 \to x_2\}$

and

(7.2) $\qquad T = \{(x_1,x_2) : x_1 \in B_1,\ x_2 \in B_2,\ \text{and}\ x_1 \to x_2\}.$

Set

$$H_1 = \{(i,j) \in C : h(z',n') \in T - S \text{ for some } (z,n,z',n') \in T - S\}$$

and

$$H_2 = \{(i,j) \in C : v' \overset{S}{\sim} (z,n,z',n'),\ (z,n,z',n') \overset{S}{\sim} v''\ \text{and}\ h(z',n') = (i,j)$$

$$\text{for some } (z,n,z',n') \in F - (T \cup S),\ v' \in T,\ \text{and}\ v'' \in S\}.$$

The set $H = H_1 \cup H_2$ is the set of all (center, class) pairs $(i,j) \in C$ such that a marked job can be of class j at center i when the passage time specified by the sets A_1, A_2, B_1, and B_2 terminates or is not underway. The labelled jobs method applies to passage times for which $S \cap T = \emptyset$ so that the set H is nonempty. (Recall that we write $h(z,n) = (i,j)$ when the job in position n in the job stack associated with state $z \in D$ is of class j at center i. We write $v' \overset{S}{\sim} v''$ when for some $n \geq 1$ there is a positive probability, starting from state v', of entering v'' on the nth step under the restriction that none of the states in S is entered in between.)

Definition 7.3. An element $z \in D$ is called a *single state of the job stack process for the passage time specified by the sets* A_1, A_2, B_1, and B_2 if

(i) $h(z,n) = (i_0, j_{l_n}(i_0)) \in H$ for some single server center i_0, $n = 1,2,\ldots,N$; and

(ii) there exists $(z_1, m) \in B_1$ such that $(z_1, m) \to (z,n)$ for some $(z,n) \in B_2$.

According to this definition a single state, z, of the job stack process for the specified passage time corresponds to a configuration of the job stack such that no passage times are underway ($h(z,n) \in H$ for all n), all jobs are at the same center (i_0) with exactly one job in service, and there exists z_1 such that a passage time for some job terminates when the job stack process jumps from z_1 to z. (Recall that $j_1(i_0), j_2(i_0), \ldots, j_{k(i_0)}(i_0)$ are the classes of jobs serviced at center i_0; the definition requires that $j_{l_n}(i_0)$, the class at center i_0 to which the job in position n belongs, satisfy $j_{l_n}(i_0) \in \{j_1(i_0), j_2(i_0), \ldots, j_{k(i_0)}(i_0)\}$ for all n.) We assume that a single state exists in the sense of Definition 7.3 and that the initial state of the job stack process is a single state, z_0.

4.8 The Underlying Stochastic Structure

The labelled jobs method rests on the result that the job stack process observed at the terminations of passage times is a regenerative process in discrete time. To show this we use a representation of the job stack process (restricted to the set D) as an irreducible GSMP with unit speeds.

Recall from Section 4.2 that the GSMP associated with the job stack process Z has state space, D, and event set, E, where event $e =$ "completion of service to a job of class j by server m at center i" $\in E(z)$ if and only if (center, class) pair $(i,j) \in U(z)$ and in state z a job of class j the class of job receives service from server m at center i. Jumps of the process from state z triggered by event e are governed by the probability mass function $q(\cdot;z,e)$. At a jump from state z to state z' triggered by event e, new clock values are generated for each event $e' \in E(z') - (E(z) - \{e\})$. The distribution function of such a new clock time (a service time for a job class at some center) is denoted by $F(\cdot;z',e',z,e)$; by assumption, each has finite mean and a density function which is continuous and positive on $(0, +\infty)$. For $e' \in E(z') \cap (E(z) - \{e\})$, the old clock reading is kept after the jump. For $e' \in (E(z) - \{e\}) - E(z')$, event z ceases to be scheduled after the jump. Denote the set of all possible events that can occur by $E = \{e_1, e_2, ..., e_M\}$.

With each $z \in D$, associate the set of clock readings

$$C(z) = \{(c_1, c_2, .., c_e) : c_i \geq 0 \text{ and } c_i > 0 \text{ if and only if } e_i \in E(z)\},$$

where c_i is the reading on the clock corresponding to event $e_i \in E(z)$. It follows from Proposition 1.3 that there exists a finite sequence $z_1, z_2, ..., z_n$ of states of the job stack process and events $e_{i_0}, e_{i_1}, ..., e_{i_n}$ satisfying $q(z_1; z, e_{i_0}) q(z_2; z_1, e_{i_1}) \cdots q(z'; z_n, e_{i_n}) > 0$ for all $z, z' \in D$.

Label the jobs from 1 to N and denote by $N^j(t)$ the position of job j in the job stack at time t, $j = 1, 2, ..., N$. Then in terms of the vector $Z(t)$ of (1.1), set

$$X^0(t) = (Z(t), N^1(t), ..., N^N(t))$$

and call the process $X^0 = \{X^0(t) : t \geq 0\}$ the *fully augmented job stack process*. Denote the state space of this process by G^0. Let z_0 be a single state of the job stack process as in Definition 7.3. Select $(z_0, n^1, ..., n^N) \in G^0$ and set $X^0(0) = (z_0, n^1, ..., n^N)$. Let $\{P_n^0 : n \geq 1\}$ be the successive passage times (irrespective of job identity) in termination order. Also let $\{T_n^0 : n \geq 1\}$ be the corresponding sequence of termination times and set $T_0^0 = 0$.

Proposition 8.2 asserts that the job stack process returns to a single state infinitely often with probability one. This leads to the result that $\{(Z(T_n^0), P_{n+1}^0) : n \geq 0\}$ is a regenerative process in discrete time. Let $M(t)$ be the last state occupied by the job stack process Z before jumping to the state occupied at time t, set $Y(t) = (M(t), Z(t))$ and denote the jump times of $Y = \{Y(t) : t \geq 0\}$ by $\{\zeta_k : k \geq 0\}$. Then by Proposition 2.2,

(8.1) $$P\{Y(\zeta_k) = (z, z') \text{ i.o.}\} = 1$$

for all $z, z' \in D$ such that $z \to z'$.

Proposition 8.2. Let z be a single state of the job stack process with respect to the passage time specified by the sets A_1, A_2, B_1, and

B_2. Suppose that the routing matrix P is irreducible and for some service center, i_0, either $k(i_0) = 1$ or service to a job of class $j_{k(i_0)}(i_0)$ at center i_0 is preempted when any other job of higher priority joins the queue. Then

(8.3) $$P\{Z(T_n^0) = z \text{ i.o.}\} = 1.$$

Proof. Condition (ii) of Definition 7.3 implies that there exists z_1 with $z_1 \to z$ such that a passage time for some job terminates when the job stack process Z jumps from state z_1 to state z. Therefore every jump time ζ_k at which $Y(\zeta_k) = (z_1, z)$ is an element of $\{T_n^0 : n \geq 0\}$ and is such that $Z(\zeta_k) = z$. The result now follows from (8.1). □

Proposition 8.4. Under the conditions of Proposition 8.2, the process $\{(Z(T_n^0), P_{n+1}^0) : n \geq 0\}$ is a regenerative process in discrete time and the expected time between regeneration points is finite.

Proof. Recall that $Z(0) = z_0$, a single state of the job stack process for the passage time. Let $\{T_{\beta_k}^0 : k \geq 1\}$ be the successive times at which a passage time terminates with the job stack process in state z_0, and set $T_{\beta_0}^0 = 0$. We must show that

(i) $\{\beta_k : k \geq 0\}$ is a renewal process in discrete time

and that for any $i_1 < i_2 < \ldots < i_m$ ($m \geq 1$) and $k \geq 0$

(ii) $\{Z(T_{\beta_k+i_1}^0), P_{\beta_k+i_1+1}^0, \ldots, Z(T_{\beta_k+i_m}^0), P_{\beta_k+i_m+1}^0\}$ and $\{Z(T_{i_1}^0), P_{i_1+1}^0, \ldots, Z(T_{i_m}^0), P_{i_m+1}^0\}$ have the same distribution, and $\{Z(T_{\beta_k+i_1}^0), P_{\beta_k+i_1+1}^0, \ldots, Z(T_{\beta_k+i_m}^0), P_{\beta_k+i_m+1}^0\}$ is independent of $\{(Z(T_n^0), P(T_{n+1}^0)) : 0 \leq n < \beta_k\}$.

Set $\zeta(\beta_k) = \min\{\zeta_n : \zeta_n > T_{\beta_k}^0\}$. At time $T_{\beta_k}^0$, no passage times are underway. Now observe that each of the clocks running at time $\zeta(\beta_k)+$ was set at time $\zeta(\beta_k)$. Therefore $\{Z(t) : t \geq \zeta(\beta_k)\}$

4.8 The Underlying Stochastic Structure   179

determines the finite dimensional distributions of $Z(T^0_{\beta_k+i})$, $P^0_{\beta_k+i+1}$ for $i \geq 1$ and the distribution of $\beta_{k+1} - \beta_k$. Now observe that each of the clocks running at time $T^0_{\beta_k}$ was set or probabilistically reset at time $T^0_{\beta_k}$. The joint distribution of $\{Z(t): t \geq \zeta(\beta_k)\}$ and the clocks set at time $\zeta(\beta_k)$ depends on the past history of the job stack process Z only through z_0 and the event occurring at time $\zeta(\beta_k)$. This distribution is the same for all β_k and therefore (i) and (ii) hold. Since $E\{T^0_{\beta_k+1} - T^0_{\beta_k}\} < \infty$ by Proposition 8.2, the state space of the process Z is finite, and the clock setting distributions have finite mean, $E\{\beta_{k+1} - \beta_k\} < \infty$. □

Note that the successive times at which $\{Z(T^0_k): k \geq 0\}$ hits state z_0 are *not* regeneration points for the job stack process Z.

Example 8.5 shows that even if the set of states of the job stack process that are accessible from a fixed single state is irreducible, the set $\{(z, n^1, n^2, \ldots, n^N) \in G^0 : z \in D\}$ of states of the fully augmented job stack process need not be irreducible.

Example 8.5. (Cyclic Queues) Let $Z(t)$ be the number of jobs waiting or in service at center 1 at time t. Consider the passage time that starts when a job enters the center 1 queue upon completion of service at center 2 and terminates when the job next joins the center 2 queue.

The sets A_1 and A_2 that define the starts of the passage time are

$$A_1 = \{(i,N): 0 \leq i < N\}$$

and

$$A_2 = \{(i,1): 0 < i \leq N\}.$$

The sets B_1 and B_2 that define the terminations of the passage time are

$$B_1 = \{(i,i): 0 < i \leq N\}$$

and

$$B_2 = \{(i-1, i) : 0 < i \le N\}.$$

State 0 is a single state of the job stack process with respect to the specified passage time. The set, D, of all states of the job stack process that are accessible from state 0 is the set $D = D^* = \{0, 1, \ldots, N\}$ and all pairs of states in the set $G = \{(z, n) \in G^* : z \in D\}$ communicate. When there are $N = 3$ jobs in the network, the fully augmented job stack process has two ireducible, closed sets of states. Observe that all pairs of states in the set $\{(z, 1, 2, 3), (z, 2, 3, 1), (z, 3, 1, 2) : z = 0, 1, 2, 3\}$ communicate as do all pairs of states in the set $\{(z, 1, 3, 2), (z, 3, 2, 1), (z, 2, 1, 3) : z = 0, 1, 2, 3\}$. For each such set there is a sequence of passage times (irrespective of job identity and enumerated in order of passage time termination).

Proposition 8.6 asserts that the sequences of passage times associated with starting the job stack process in a fixed single state converge to a common random variable. The proof is similar to that of Proposition 2.1 in Appendix 2.

Proposition 8.6. Suppose that x and x' communicate for all $x, x' \in G$. Let z_0 be a single state of the job stack process for the passage time specified by the sets $A_1, A_2, B_1,$ and B_2. Denote $X^0(0)$ by $(z_0, n^1, \ldots, n^N)$ and let $\{P_n^0 : n \ge 1\}$ be the successive passage times (irrespective of job identity) enumerated in termination order. Then $P_n^0 \Rightarrow P^0$ as $n \to \infty$ for all $(z_0, n^1, \ldots, n^N) \in G^0$, where P^0 is the limiting passage time for any marked job.

4.9 The Labelled Jobs Method

The goal of the simulation is the estimation of $r^0(f) = E\{f(P^0)\}$, where f is a real-valued (measurable) function and P^0 is the limiting passage time for any marked job. To obtain estimates for $r^0(f)$, select a single state, z_0, of the job stack process (for the passage time specified by the sets $A_1, A_2, B_1,$ and B_2) and an initial state $(z_0, n^1, \ldots, n^N)$ of the fully augmented job stack process X^0. Carry out the simulation of the fully augmented job stack process in blocks defined by the successive times $\{T^0_{\beta_k}: k \geq 1\}$ at which a passage time terminates with the job stack process in state z_0 ($\beta_0 = 0$ and $T^0_{\beta_0} = 0$). Set

$$Y^0_m(f) = \sum_{j=\beta_{m-1}+1}^{\beta_m} f(P^0_j)$$

and $\alpha^0_m = \beta_m - \beta_{m-1}$, $m \geq 1$. Propositions 9.1 and 9.2 follow from Proposition 8.4 and general results for regenerative processes.

Proposition 9.1. The sequence of pairs of random variables $\{(Y^0_m(f), \alpha^0_m) : m \geq 1\}$ are independent and identically distributed.

Proposition 9.2. Provided that $P\{P^0 \in D(f)\} = 0$ and $E\{|f(P^0)|\} < \infty$,

$$E\{f(P^0)\} = \frac{E\{Y^0_1(f)\}}{E\{\alpha^0_1\}}.$$

With these results

$$\hat{r}^0(n) = \frac{\bar{Y}^0(n)}{\bar{\alpha}^0(n)} = \frac{\sum_{m=1}^{n} Y^0_m(f)}{\sum_{m=1}^{n} \alpha^0_m}$$

is a strongly consistent point estimate for $r^0(f)$ and an asymptotic $100(1 - 2\gamma)\%$ confidence interval is

$$\left[\hat{r}^0(n) - \frac{z_{1-\gamma} \, s^0(n)}{\bar{\alpha}^0(n) \, n^{1/2}}, \; \hat{r}^0(n) + \frac{z_{1-\gamma} \, s^0(n)}{\bar{\alpha}^0(n) \, n^{1/2}} \right].$$

The quantity $(s^0(n))^2$ is a strongly consistent point estimate for $(\sigma^0(f))^2 = \text{var}(Y^0_1(f) - r^0(f)\alpha^0_1)$. Asymptotic confidence intervals for $r^0(f)$ are based on the c.l.t.

(9.3) $$\frac{n^{1/2}\{\hat{r}^0(n) - r^0(f)\}}{\sigma^0(f)/E\{\alpha_1^0\}} \Rightarrow N(0,1)$$

as $n \to \infty$.

Algorithm 9.4. (Labelled Jobs Method for Non-Markovian Networks)

1. Select a single state, z_0, of the job stack process (so that there exists z_1 such that a passage time for some job terminates when the job stack process makes a transition from state z_1 to state z_0) and an initial state $(z_0, n^1, \ldots, n^N)$ for the fully augmented job stack process.

2. Set $X^0(0) = (z_0, n^1, \ldots, n^N)$ and simulate the fully augmented job stack process. Observe a fixed number, n, of blocks defined by the successive times $\{T_{\beta_k}^0 : k \geq 1\}$ at which a passage time terminates and the job stack process hits state z_0. In each block measure all passage times for all of the jobs.

3. Compute the number, α_m^0, of passage times in the mth block, and the quantity

$$Y_m^0(f) = \sum_{j=\beta_{m-1}+1}^{\beta_m} f(P_j^0).$$

4. Form the point estimate

$$\hat{r}^0(n) = \frac{\bar{Y}^0(n)}{\bar{\alpha}^0(n)}.$$

7. Form the asymptotic $100(1 - 2\gamma)\%$ confidence interval

$$\left[\hat{r}^0(n) - \frac{z_{1-\gamma} s^0(n)}{\bar{\alpha}^0(n) n^{1/2}}, \hat{r}^0(n) + \frac{z_{1-\gamma} s^0(n)}{\bar{\alpha}^0(n) n^{1/2}}\right].$$

Example 9.5. (Cyclic Queues With Nonpreemptive Priority) As center 1 sees only one job class, we can take $i_0 = 1$ so that with $N = 3$ jobs, $z_1^* = (3,0,0,0)$. Since all states of the job stack process defined by (5.9) except $(0,0,2,2)$ and $(0,1,1,1)$ are accessible from state z_1^*, the set $D = D^* - \{(0,0,2,2),(0,1,1,1)\}$. By Proposition 1.3, any two states $z,z' \in D$ communicate.

Consider the passage time specified by the subsets A_1, A_2, B_1, and B_2 of the set G:

$$A_1 = \{(1,0,1,2,1),(1,1,0,1,1),(2,0,0,0,2)\},$$

$$A_2 = \{(0,1,1,2,1),(0,2,0,1,1),(1,1,0,1,2)\},$$

$$B_1 = \{(0,1,1,2,2),(1,0,1,2,2)\},$$

and

$$B_2 = \{(1,1,0,1,1),(2,0,0,0,1),(1,0,1,2,1)\}.$$

This passage time starts when a job joins the queue at center 2 as class 1 and terminates when the job completes service at center 2 as class 2. The sets S and T defined by (7.1) and (7.2) satisfy $S \cap T = \emptyset$. (Note that because $(0,0,2,2) \notin D$, $(0,0,2,2,2) \notin B_1$.) For this passage time, $H = \{(1,1)\}$ and $z_0 = (3,0,0,0)$ is a single state. Clearly, z_0 satisfies condition (i) of Definition 7.3. Condition (ii) is also satisfied because $(1,0,1,2,2) \rightarrow (2,0,0,0,1)$.

Now consider the passage time specified by the subsets A_1, A_2, B_1, and B_2 of G:

$$A_1 = \{(1,0,1,2,2),(0,0,2,2,2),(0,1,1,2,2)\},$$

$$A_2 = \{(2,0,0,1,1),(1,0,1,2,1),(1,1,0,1,1)\},$$

$$B_1 = \{(1,1,0,1,2),(0,2,0,1,2)\},$$

and

$$B_2 = \{(1,0,1,2,2),(0,1,1,2,2)\}.$$

The passage time starts when a job completes service at center 2 as class 2 and terminates when the job completes service at center 2 as class 1. For this passage time, the set $H = \{(2,2)\}$. There is at least one passage time underway unless the configuration of the job stack is $(0,0,2,2)$. As $(0,0,2,2) \notin D$, there is no single state and the labelled jobs method does not apply.

4.10 Comparison of Methods

The marked job method provides point and interval estimates for general passage times and prescribes observation of passage times for an arbitrarily chosen, distinguished job. The half length of the confidence interval (obtained from a simulation of fixed length) for the expected value of a general function f of the limiting passage time is proportional to a certain quantity $e(f)$. The labelled jobs method provides estimates for passage times through a subnetwork and observed passage times for all the jobs are used to construct point and interval estimates. Using the labelled jobs method (with the same constant of proportionality) the half length of the confidence interval is proportional to a quantity $e^0(f)$. Since the quantities $e(f)$ and $e^0(f)$ are independent of the blocks of the underlying regenerative process, they are appropriate measures of the statistical efficiency of the estimation procedures.

For closed networks of queues with exponential service times it is possible to compute theoretical values for expected passage times and the associated variance constants appearing in c.l.t.'s used to form confidence intervals. This leads to a quantitative assessment of the relative statistical efficiency of the marked job and labelled jobs methods for Markovian networks. For networks of queues with general service times, there is little hope of computing the needed theoretical values, even for expected passage times. Using central limit theorem and continuous mapping theorem arguments, in this section we show that $e^0(f) \leq e(f)$ for any function f (and all numbers of jobs in the network) so that confidence intervals constructed using the labelled jobs method are shorter than those obtained from the marked job method. This is consistent with intuition since the labelled jobs method extracts more passage time information from a fixed length simulation run.

Assume that for the passage time P the set S' of (4.8) is nonempty, the sets S and T of (7.1) and (7.2) are disjoint, and there exists a single state of the job stack process in the sense of Definition 7.3. (These assumptions ensure that both the marked job and labelled jobs methods are applicable.) Let $L(t)$ be the last state occupied by the augmented job stack process X before jumping to $X(t)$. Similarly, let $L^0(t)$ be the last state occupied by the fully augmented job stack process X^0 before jumping to $X^0(t)$. Set $V(t) = (L(t), X(t))$ and $V^0(t) = (L^0(t), X^0(t))$. Denote the state space of $V = \{V(t): t \geq 0\}$ by F and the state space of $V^0 = \{V^0(t): t \geq 0\}$ by F^0. We work with the process V^0 and take job 1 to be the marked job. Select a single state, z_0, of the job stack process with respect to the specified passage time. Also select $v_0 = (z_0', m_1', z_0, m_1) \in T$, and

$$v_0^0 = (z_0', m_1', m_2', \ldots, m_N', z_0, m_1, m_2, \ldots, m_N) \in F^0.$$

Take $V^0(0) = v_0^0$.

Lemma 10.1 is the basis for construction of c.l.t.'s in continuous time for sequences of passage times. Let $\{\gamma_k^*: k \geq 0\}$ be a renewal process defined on the same sample space as $\{P_n^0: n \geq 1\}$ and let $\{P_{n_j}^0: j \geq 1\}$ be a subsequence of $\{P_n^0: n \geq 1\}$. Set $\delta_{k+1}^* = \gamma_{k+1}^* - \gamma_k^*$ and denote by $m^*(t)$ the number of $P_{n_j}^0$ completed in $(0, t]$. Also set $\alpha_k^* = m^*(\gamma_k^*) - m^*(\gamma_{k-1}^*)$ and

$$Y_k^*(f) = \sum_{j=m^*(\gamma_{k-1}^*)+1}^{m^*(\gamma_k^*)} f(P_{n_j}^0).$$

Lemma 10.1. Suppose that the pairs of random variables $\{(Y_k^*(f), \alpha_k^*) : k \geq 1\}$ are independent and identically distributed and

$$r(f) = \frac{E\{Y_1^*(f)\}}{E\{\alpha_1^*\}}.$$

Also suppose that $(\sigma^*(f))^2 \equiv \text{var}(Y_1^*(f) - r(f)\alpha_1^*) < \infty$. Then provided that $E\{\delta_1^*\} < \infty$, $E\{(\alpha_1^*)^2\} < \infty$, and $E\{(Y_1^*(|f|))^2\} < \infty$,

$$\frac{t^{1/2}\left(\frac{1}{m^*(t)} \sum_{j=1}^{m^*(t)} f(P_{n_j}^0) - r(f)\right)}{(E\{\delta_1^*\})^{1/2} \sigma^*(f)/E\{\alpha_1^*\}} \Rightarrow N(0,1)$$

as $t \to \infty$.

Proof. Set

$$A(t) = \sum_{j=1}^{m^*(t)} f(P_{n_j}^0) - r(f) m^*(t)$$

and let $\{n^*(t) : t \geq 0\}$ be the counting process associated with the renewal process $\{\gamma_k^* : k \geq 0\}$. Then the Doeblin decomposition is given by

(10.2) $$A(t) = \sum_{k=1}^{n^*(t)} [Y_k^*(f) - r(f)\alpha_k^*]$$

$$+ \sum_{j=m^*(\gamma_{n^*(t)}^*)+1}^{m^*(t)} f(P_{n_j}^0) - r(f)[m^*(t) - m^*(\gamma_{n^*(t)}^*)].$$

Also set $Z_k^*(f) = Y_k^*(f) - r(f)\alpha_k^*$. By assumption, $E\{Z_k^*(f)\} = 0$ and the $Z_k^*(f)$'s are i.i.d. The weak law for renewal counting processes yields $n^*(t)/t \Rightarrow 1/E\{\delta_1^*\}$ as $t \to \infty$. These facts, together with the c.l.t. for random partial sums, and the continuous mapping theorem, yields the c.l.t.

(10.3) $$\frac{\sum_{k=1}^{n^*(t)} Z_k^*(f)}{\sigma^*(f)\, t^{1/2}/\left(E\{\delta_1^*\}\right)^{1/2}} \Rightarrow N(0,1)$$

as $t \to \infty$. It follows from (10.3) that the first term (suitably normalized) on the right hand side of (10.2) converges weakly to $N(0,1)$. The last two terms on the right hand side of (10.2) are remainder terms that, when divided by $t^{1/2}$, converge weakly to 0. This argument is standard but requires $E\{(\alpha_1^*)^2\}$ and $E\{(Y_1^*(|f|))^2\}$ to be finite. The converging together lemma and (10.3) ensure that

(10.4) $$\frac{A(t)}{\sigma^*(f)\, t^{1/2}/\left(E\{\delta_1^*\}\right)^{1/2}} \Rightarrow N(0,1)$$

as $t \to \infty$. Finally, dividing the numerator and denominator of the left hand side of (10.4) by $m^*(t)$, the desired result follows from the continuous mapping theorem and the fact that

$$\frac{m^*(t)}{t} \Rightarrow \frac{E\{\alpha_1^*\}}{E\{\delta_1^*\}}$$

as $t \to \infty$. □

The labelled jobs method prescribes selection of z_0, a single state of the job stack process with respect to the specified passage time, along with $v_0 = (z_0', m_1', z_0, m_1) \in T$ and simulation of the process V^0 in blocks defined by the exits from the set $\{(z_0', n_1', n_2', \ldots, n_N', z_0, n_1, n_2, \ldots, n_N) \in F^0 : (z_0', n_i', z_0, n_i) = v_0 \text{ for some } i, 1 \le i \le N\}$. Let δ_1^0 be the length of a block and $m^0(t)$ be the number of passage times (irrespective of job identity) completed in the interval $(0, t]$. Set $(\sigma^0(f))^2 = \text{var}(Y_1^0(f) - r(f)\alpha_1^0)$. Then, by

188 4 Non-Markovian Networks

Lemma 10.1,

$$\frac{t^{1/2}\left(\frac{1}{m^0(t)}\sum_{n=1}^{m^0(t)} f(P_n^0) - r(f)\right)}{(E\{\delta_1^0\})^{1/2} \sigma^0(f)/E\{\alpha_1^0\}} \Rightarrow N(0,1)$$

as $t \to \infty$. This c.l.t. implies that the half length of the confidence interval obtained from a simulation of fixed length is proportional to

(10.5) $\qquad e^0(f) = (E\{\delta_1^0\})^{1/2} \sigma^0(f)/E\{\alpha_1^0\},$

(We assume that $E\{(\alpha_1^0)^2\} < \infty$ and $E\{(Y_1^0(|f|))^2\} < \infty$. This makes it possible to apply Lemma 10.1.) Since the numerator in this c.l.t. and the limit ($N(0,1)$) is independent of the state v_0 selected from T, so is the denominator; this is a consequence of the convergence of types theorem. Thus, $e^0(f)$ is an appropriate measure of the statistical efficiency of the labelled jobs method.

The marked job method prescribes selection of $x' = (z',n') \in A_2'$. Simulation of the process V is in blocks defined by the successive entrances to the set $\{(z_0, N, z^*, n^*) \in S' : z^* = z', n^* = n'\}$. Let δ_1 be the length of a block and $m(t)$ be the number of passage times completed in the interval $(0, t, rb$. Set $\sigma^2(f) = \text{var}(Y_1(f) - r(f)\alpha_1)$. Then, by Lemma 10.1,

(10.6) $\qquad \dfrac{t^{1/2}\left(\frac{1}{m(t)}\sum_{n=1}^{m(t)} f(P_n) - r(f)\right)}{(E\{\delta_1\})^{1/2} \sigma(f)/E\{\alpha_1\}} \Rightarrow N(0,1)$

as $t \to \infty$. Using an argument similar to that used for the labelled jobs method, an appropriate measure of the statistical efficiency of the marked job method is the quantity

(10.7) $\qquad e(f) = (E\{\delta_1\})^{1/2} \sigma(f)/E\{\alpha_1\}.$

Proposition 10.12 asserts that $e^0(f) \leq e(f)$ for all functions f.

In terms of the single state, z_0, and the state $v_0 = (z'_0, m'_1, z_0, m_1) \in T$, define subsets $U^1, U^2, \ldots, U^N$ of F^0 by

$$U^i = \{(z'_0, m'_1, \ldots, m'_i, n'_{i+1}, \ldots, n'_N, z_0, m_1, \ldots, m_i, n_{i+1}, \ldots, n_N) \in F^0\},$$

$1 \le i \le N$. According to this definition, entrance of the process V^0 to the set U^i corresponds to termination of a passage time for job 1 (in position m_1 of the job stack associated with state z_0) with job $2, \ldots, i$ in fixed positions $m_2, \ldots, m_i$ of the job stack associated with state z_0. Observe that the times at which the process V^0 hits the set U^i are a subsequence of the times at which V^0 hits the set U^{i-1}. Comparison of the marked job and labelled jobs methods rests on c.l.t.'s in continuous time for U^i blocks (defined by the successive exits from the set U^i) of V^0.

Denote by δ_k^i the length of the kth U^i block of V^0, $k \ge 1$. Let α_k^{ji} be the number of passage times for job j in the kth U^i block and let Y_k^{ji} be the sum of the values of the function f for these passage times, $1 \le j \le i$. Then set

$$e^{ji} = \left(E\{\delta_1^i\}\right)^{1/2} \sigma^{ji} / E\{\alpha_1^{ji}\},$$

where $(\sigma^{ji})^2 = \mathrm{var}\,(Y_k^{ji} - r(f)\alpha_k^{ji})$. Denote the analogous quantities for jobs $1, 2, \ldots, j$ by $\alpha_k^{(j)i}$, $Y_k^{(j)i}$, and

$$e^{(j)i} = \left(E\{\delta_1^{(j)i}\}\right)^{1/2} \sigma^{(j)i} / E\{\alpha_1^{(j)i}\},$$

respectively, where $(\sigma^{(j)i})^2 = \mathrm{var}\,(Y_k^{(j)i} - r(f)\alpha_k^{(j)i})$.

We require three lemmas. The first asserts that the statistical efficiency of a simulation based on observation of passage times for job 1 (the marked job) in U^1 blocks is the same as for a simulation based on observation of passage times for job i in U^i blocks, $1 \le i \le N$.

Lemma 10.8. $e^{ii} = e^{11}$.

Proof. Recall that $N^i(t)$ is the position of the job labelled i in the job stack at time t. Set $X^i(t) = (Z(t), N^i(t))$ and consider the process

$X^i = \{X^i(t): t \geq 0\}$. Let $L^i(t)$ be the last state occupied by X^i before jumping to $X^i(t)$ and set $V^i(t) = (L^i(t), X^i(t))$. Let $\{P_n^i : n \geq 0\}$ be the sequence of passage time for job i enumerated in termination order. Observe that the successive entrances to $v_0 = (z_0', m_1', z_0, m_1)$ are regeneration points for the process $V^i = \{V^i(t): t \geq 0\}$. Let $m^i(t)$ be the number of passage times for job i completed in $(0,t]$. Based on these v_0-cycles (for some e^i) by Lemma 10.1

$$\frac{t^{1/2}\left(\dfrac{1}{m^i(t)} \sum_{n=1}^{m^i(t)} f(P_n^i) - r(f)\right)}{e^i} \Rightarrow N(0,1)$$

as $t \to \infty$. Similarly, based on U^i blocks of V^i,

(10.9) $$\frac{t^{1/2}\left(\dfrac{1}{m^i(t)} \sum_{n=1}^{m^i(t)} f(P_n^i) - r(f)\right)}{e^{ii}} \Rightarrow N(0,1)$$

as $t \to \infty$. Since the numerators and the limits are the same, the denominators must be the same: $e^{ii} = e^i$.

To complete the proof, we show that $e^i = e^{11}$. Set

$$\zeta_1^i = \inf\{t > 0 : V^i(t) = v_0\}$$

and $W^i(t) = V^i(\zeta_1^i + t)$. The process V is a GSMP as is the process $W^i = \{W^i(t): t \geq 0\}$. These two processes have the same probability mass functions and clock setting distributions since these building blocks are functions only of the routing matrix and the parameters of the service time distributions, and the jobs are stochastically identical. This implies that V and W^i have the same finite dimensional distributions. Enumerate the passage times for job i from time ζ_1^i as $\{R_n^i : n \geq 0\}$ and let $m^i(t)$ be the number of passage times for job i completed in $(\zeta_1^i, \zeta_1^i + t]$. Denote the sequence of passage times for job 1 by $\{P_n^1 : n \geq 0\}$ and let $m^1(t)$ be the number of passage times for job 1 completed in $(0,t]$. Now set

$$C^1(t) = \frac{1}{m^1(t)} \sum_{n=1}^{m^1(t)} f(P_n^1) - r(f)$$

and consider the functional $\{C^1(t): t \geq 0\}$. Also set

$$C^1(t) = \frac{1}{m^1(t)} \sum_{n=1}^{m^1(t)} f(P_n^1) - r(f).$$

and observe that $\{C^i(t): t \geq 0\}$ is the same functional of the process W^i so that for all $t > 0$, $C^1(t)$ and $C^i(t)$ have the same distribution. Lemma 10.1 implies that

$$\frac{t^{1/2} C^1(t)}{e^{11}} \Rightarrow N(0,1)$$

and

$$\frac{t^{1/2} C^i(t)}{e^i} \Rightarrow N(0,1)$$

as $t \to \infty$; therefore $e^i = e^{11}$. □

Lemma 10.10 asserts that simulations based on observation of passage times for jobs $1,2,...,j$ in U^i blocks and in U^{i+1} blocks are equally efficient.

Lemma 10.10. Let $l = i + 1$. Then $e^{(j)i} = e^{(j)l}$ for $j = 1,2,...,i$.

Proof. Set $v_0^{(k)} = (z_0', m_1', m_2', ..., m_k', z_0, m_1, m_2, ..., m_k)$ and $X^{(k)}(t) = (Z(t), N^1(t), ..., N^k(t))$, $1 \leq k \leq N$. Let $L^{(k)}(t)$ be the last state occupied by the process $X^{(k)} = \{X^k(t): t \geq 0\}$ before jumping to $X^{(k)}(t)$. Also set $V^{(k)}(t) = (L^{(k)}(t), X^{(k)}(t))$. The successive entrances to state $v_0^{(i)}$ are regeneration points for the process $V^{(i)} = \{V^i(t): t \geq 0\}$ and the successive entrances to state $v_0^{(l)}$ are regeneration points for the process $V^{(l)}$. Let $\{P_n^{(j)}: n \geq 0\}$ be the sequence of passage times for jobs $1,2,...,j$ enumerated in termination order and let $m^{(j)}(t)$ be the number of passage times for jobs 1,2,...,*j* completed in (0,*t*]. Then

$$\frac{t^{1/2}\left(\frac{1}{m^{(j)}(t)}\sum_{n=1}^{m^{(j)}(t)}f(P_n^{(j)})-r(f)\right)}{e^{(j)i}}\Rightarrow N(0,1)$$

and

$$\frac{t^{1/2}\left(\frac{1}{m^{(j)}(t)}\sum_{n=1}^{m^{(j)}(t)}f(P_n^{(j)})-r(f)\right)}{e^{(j)l}}\Rightarrow N(0,1)$$

as $t\to\infty$. Since the numerators are the limits are the same, the denominators must be the same. □

Lemma 10.11 asserts that a simulation based on observation of passage times for jobs 1 and 2 in U^2 blocks is at least as efficient as a simulation based on passage times for job 1 in U^1 blocks. The idea is to develop a bivariate c.l.t. for passage times of jobs 1 and 2 in U^2 blocks, apply the continuous mapping theorem to obtain a one dimensional c.l.t., and make an identification of the resulting variance constant.

Lemma 10.11. $e^{(2)2} \leq e^{11}$.

Proof. Set $W_k^1 = Y_k^{12} - r(f)\alpha_k^{12}$ and $W_k^2 = Y_k^{22} - r(f)\alpha_k^{22}$. Let W_k be the column vector with components W_k^1 and W_k^2. Also set $\Sigma = \{\sigma_{ij}\}$ and $\sigma_{ij} = E\{W_1^i W_1^j\}$ for $i,j = 1,2$. The vectors $\{W_k : k \geq 0\}$ are i.i.d. with mean **0** and finite variance, and it follows that

$$\frac{1}{n^{1/2}}\sum_{k=1}^{n}W_k \Rightarrow N(0,\Sigma),$$

as $n \to \infty$. Set

$$\sigma^2 = \sum_i \sigma_{ii} + 2 \sum_{i<j} \sigma_{ij}.$$

An application of the continuous mapping theorem using the mapping $h(w_1, w_2) = w_1 + w_2$ yields

$$\frac{1}{\sigma n^{1/2}} \sum_{k=1}^{n} \sum_{i=1}^{2} W_k^i \Rightarrow N(0,1),$$

as $n \to \infty$. Furthermore from the form of this c.l.t. (since $W_k^1 + W_k^2$ corresponds to jobs 1 and 2), it is clear that $\sigma^2 = (\sigma^{(2)2})^2$, $\sigma_{11} = (\sigma^{12})^2$, and $\sigma_{22} = (\sigma^{22})^2$.

Next observe that $e^{22} = e^{11}$ by Lemma 10.8 and $e^{12} = e^{11}$ by Lemma 10.10. Using the definitions of e^{12}, e^{11}, and e^{22}, these results imply

$$\sigma_{11} = \frac{(\sigma^{11})^2 E\{\delta_1^1\}(E\{\alpha_1^{12}\})^2}{E\{\delta_1^2\}(E\{\alpha_1^{11}\})^2}$$

and

$$\sigma_{22} = \frac{(\sigma^{11})^2 E\{\delta_1^1\}(E\{\alpha_1^{22}\})^2}{E\{\delta_1^2\}(E\{\alpha_1^{11}\})^2}.$$

From the Cauchy-Schwarz inequality we have

$$2|\sigma_{12}| \leq 2(\sigma_{11}\sigma_{22})^{1/2} = \frac{(\sigma^{11})^2 E\{\delta_1^1\}(2E\{\alpha_1^{22}\}E\{\alpha_1^{12}\})}{E\{\delta_1^2\}(E\{\alpha_1^{11}\})^2}.$$

Hence

$$(\sigma^{(2)2})^2 = \sigma_{11} + \sigma_{22} + 2\sigma_{12}$$
$$\leq \frac{(\sigma^{11})^2 E\{\delta_1^1\}(E\{\alpha_1^{22}\} + E\{\alpha_1^{12}\})^2}{E\{\delta_1^2\}(E\{\alpha_1^{11}\})^2}.$$

Since
$$E\{\alpha_1^{(2)2}\} = E\{\alpha_1^{22}\} + E\{\alpha_1^{12}\},$$

it follows that $e^{(2)2}/e^{11} \leq 1$. $\square$

Proposition 10.12. *For all functions f, $e^0(f) \leq e(f)$.*

Proof. It is sufficient to show that $e^{(N)N} \leq e^{11}$. To see this, let $m^0(t)$ be the number of passage times (irrespective of job identity) completed in $(0,t]$. Observe (again by Lemma 10.1) that

$$\frac{t^{1/2}\left(\frac{1}{m^0(t)}\sum_{n=1}^{m^0(t)} f(P_n^0) - r(f)\right)}{e^{(N)N}} \Rightarrow N(0,1)$$

as $t \to \infty$ and

$$\frac{t^{1/2}\left(\frac{1}{m^0(t)}\sum_{n=1}^{m^0(t)} f(P_n^0) - r(f)\right)}{e^0(f)} \Rightarrow N(0,1)$$

as $t \to \infty$; therefore $e^{(N)N} = e^0(f)$. But $e(f) = e^{11}$; this follows from the c.l.t.'s of (10.6) and (10.9) (with $i = 1$).

To complete the proof we show that $e^{(N)N} \leq e^{11}$. We first show that $e^{(3)3} \leq e^{11}$. Set $W_k^1 = Y_k^{(2)3} - r(f)\alpha_k^{(2)3}$ and $W_k^2 = Y_k^{33} - r(f)\alpha_k^{33}$ and let W_k be the column vector with components W_k^1 and W_k^2. Set $\Sigma = \{\sigma_{ij}\}$ and $\sigma_{ij} = E\{W_1^i W_1^j\}$. The vectors $\{W_k : k \geq 0\}$ are i.i.d. with mean $\mathbf{0}$ and therefore

$$\frac{1}{n^{1/2}}\sum_{k=1}^{n} W_k \Rightarrow N(\mathbf{0}, \Sigma),$$

4.10 Comparison of Methods 195

as $n \to \infty$. By the argument used in the proof of Lemma 10.11, $\sigma_{11} = (\sigma^{(2)3})^2$ and $\sigma_{22} = (\sigma^{33})^2$. By Lemma 10.8, $e^{33} = e^{11}$. Therefore

$$\sigma_{22} = \frac{(\sigma^{11})^2 E\{\delta_1^1\}(E\{\alpha_1^{33}\})^2}{E\{\delta_1^3\}(E\{\alpha_1^{11}\})^2}.$$

Next observe that $e^{(2)2} \le e^{11}$ by Lemma 10.11 and $e^{(2)3} = e^{(2)2}$ by Lemma 10.10. Thus,

$$\sigma_{11} \le \frac{(\sigma^{11})^2 E\{\delta_1^1\}(E\{\alpha_1^{(2)3}\})^2}{E\{\delta_1^3\}(E\{\alpha_1^{11}\})^2}.$$

By the Cauchy-Schwarz inequality

$$(\sigma^{(3)2})^2 \le \sigma_{11} + \sigma_{22} + 2(\sigma_{11}\sigma_{22})^{1/2}$$

Since $E\{\alpha_1^{(3)3}\} = E\{\alpha_1^{33}\} + E\{\alpha_1^{(2)3}\}$, it follows that $e^{(3)3}/e^{11} \le 1$. The same argument (using $e^{(3)3} \le e^{11}$) shows that $e^{(4)4} \le e^{11}$. Continuing in this way, it follows that $e^{(N)N} \le e^{11}$. □

Appendix 1

Limit Theorems for Stochastic Processes

Limit theorems pertinent to simulation output analysis involve three modes of convergence. A random variable (r.v.) is defined on a probability space $(\Omega,\mathcal{F},P)$, where Ω is the space of elementary outcomes, $\mathcal{F}$ is the family of events (subsets of Ω) for which probabilities are defined, and P is a probability measure. Unlike real numbers, sequences of r.v.'s may converge in more than one way. The strongest mode of convergence is *convergence with probability one*, also called *almost sure (a.s) convergence* or *strong convergence*.

Definition 1.1 Let $\{X_n : n \geq 1\}$ and X be r.v.'s defined on a common probability space $(\Omega,\mathcal{F},P)$. Then X_n is said to *converge with probability one* to X if

$$P\{\lim_{n \to \infty} X_n = X\} = 1$$

in which case we write $X_n \to X$ a.s. as $n \to \infty$.

An equivalent characterization of convergence with probability one is that for every $\varepsilon > 0$

$$\lim_{n \to \infty} P\{|X_m - X_n| \leq \varepsilon \text{ for all } m \geq n\} = 1.$$

Thus for all n sufficiently large ($n \geq N(\varepsilon,\delta)$), $|X_m - X_n| \leq \varepsilon$ for all $m \geq n$ with probability greater than $1 - \delta$, where $\delta > 0$ is an arbitrarily small number. The second mode of convergence is *convergence in probability*.

Definition 1.2. Let $\{X_n : n \geq 1\}$ and X be r.v.'s defined on a common probability space $(\Omega, \mathscr{F}, P)$. Then X_n is said to *converge in probability* X if for every $\varepsilon > 0$

$$\lim_{n \to \infty} P\{|X_n - X| \leq \varepsilon\} = 1$$

in which case we write $X_n \to X$ in probability as $n \to \infty$.

Thus for n sufficiently large $(n \geq N(\varepsilon, \delta))$, $|X_n - X| \leq \varepsilon$ with probability greater than $1 - \delta$. This is a statement about the distribution of $|X_n - X|$ for a single value of n, whereas convergence with probability one says something about all n beyond a certain point. Clearly, convergence with probability one implies convergence in probability.

The third mode of convergence is *convergence in distribution* (*weak convergence*). Recall that the distribution, F, of a r.v., X, satisfies $F(x) = P\{X \leq x\}$, $-\infty < x < +\infty$. Since only one distribution at a time is involved in this definition, there is no need to have $\{X_n : n \geq 1\}$ and X all defined on the same probability space. In fact, each X_n $(n \geq 1)$ and X can be defined on different probability spaces. Set $F_n(x) = P\{X_n \leq x\}$ and $F(x) = P\{X \leq x\}$.

Definition 1.3. A sequence of random variables $\{X_n : n \geq 1\}$ is said to *converge in distribution* to X if $F_n(x) \to F(x)$ for all x for which F is continuous, in which case we write $X_n \Rightarrow X$ as $n \to \infty$.

An equivalent characterization of convergence in distribution of X_n to X is

$$E\{f(X_n)\} \to E\{f(X)\}$$

for all bounded and continuous functions f. The notion of convergence in distribution does *not* require that the X_n's "settle down" as a convergent sequence of ordinary real numbers does. All three modes of convergence can be extended to random vectors in the obvious way.

The *continuous mapping theorem* (c.m.t) and the *converging together lemma* are two powerful tools that can frequently be used to derive new limit theorems from existing ones. Suppose that $\{X_n : n \geq 1\}$ is a sequence of k-dimensional random vectors ($k \geq 1$) and that X is a k-dimensional limit random vector. Let h be a function that maps $\mathbf{R}^k$ (k-dimensional Euclidean space) into the real line $\mathbf{R}^1$. Also let $D(h)$ be the set of discontinuities of the function f.

Theorem 1.4. (Continuous Mapping Theorem) Suppose that $X_n \Rightarrow X$ as $n \to \infty$. If either h is continuous or $P\{X \in D(h)\} = 0$, then $h(X_n) \Rightarrow h(X)$ as $n \to \infty$.

For the converging together lemma, let X be a limit random vector. Also let $\{X_n : n \geq 1\}$ and $\{Y_n : n \geq 1\}$ be k-dimensional random vectors and suppose that for each n the random variables X_n and Y_n are defined on the same probability space. The Euclidean norm of x is denoted by $\|x\|$.

Lemma 1.5. (Converging Together Lemma) If $X_n \Rightarrow X$ and $\|X_n - Y_n\| \Rightarrow 0$, then $Y_n \Rightarrow X$ as $n \to \infty$.

Under certain conditions, convergence in distribution of a two-dimensional random vector follows from convergence in distribution of the two components of the random vector.

Lemma 1.6. Let $\{X_n : n \geq 1\}$ and $\{Y_n : n \geq 1\}$ be two sequences of random variables defined on the same probability space. If $X_n \Rightarrow X$ and $Y_n \Rightarrow c$, then $(X_n, Y_n) \Rightarrow (X, c)$ as $n \to \infty$.

Occasionally, we deal not with a sequence, $\{X_n : n \geq 1\}$ of random variables, but with a stochastic process $\{X(t) : t \geq 0\}$ indexed by a continuous parameter, t.

Definition 1.7. The process $\{X(t): t \geq 0\}$ converge in distribution to X as $t \to \infty$ if and only if $X_{t_n} \Rightarrow X$ for every subsequence $t_n \to \infty$.

We now review the principal classes of stochastic processes used in the analysis of stochastic simulations. A stochastic process is a family of r.v.'s (or vectors) defined on a common probability space $(\Omega, \mathscr{F}, P)$. Usually the elements of this family are indexed by a time parameter, either discrete or continuous.

The most tractable stochastic process consists of an infinite sequence $\{X_n: n \geq 1\}$ of i.i.d. r.v.'s (or vectors). For this type of stochastic process all of the limit theorems of classical probability theory are available, e.g., laws of large numbers and central limit theorems. The results of classical statistics are directed to this kind of stochastic process, and when they can be applied, the analysis of simulation output is greatly simplified. This is exactly what the regenerative method (discussed in Chapter 2) is designed to do.

Markov chains generalize a stochastic process beyond an i.i.d. sequence by providing for a limited amount of dependence between the r.v.'s. Discrete time as well as continuous time Markov chains are of interest.

Definition 1.8. A *discrete time Markov chain* with finite or countable state space, S, is a sequence $\{X_n: n \geq 0\}$ of random variables taking values in S and satisfying the Markov property:

$$(1.9)\ P\{X_{n+1} = j \mid X_0 = i_0, \ldots, X_n = i\} = P\{X_{n+1} = j \mid X_n = i\} \equiv p_{ij},$$

where $i_0, i_1, \ldots, i, j \in S$.

From (1.9) it is clear that the future evolution of a DTMC depends on the past only through the present. In addition to the p_{ij}'s defined above, we need to specify the initial state (or distribution) at time 0 ($n = 0$) in order to determine the complete trajectory of the process. Set $p_i = P\{X_0 = i\}$, $i \in S$. Analytically, the motion of the DTMC depends on the one-step transition probabilities

$$P = \{p_{ij} : i, j \in S\}$$

and the initial distribution, $p = \{p_i : i \in S\}$.

A DTMC is said to be *irreducible* if any state can be reached from any other state: for all $i, j \in S$ there exists an n such that $p_{ij}^n > 0$, where p_{ij}^n is the (i,j)-entry of the nth power of the one-step transition probability matrix P. A state of a DTMC is said to be *recurrent* if starting from any state the probability of returning to the state is one; a recurrent state is *positive recurrent* if the mean time of first return to the state is finite.

The sample function of a CTMC is a piecewise constant function. The sequence of states visited is governed by a DTMC and the holding times (level portions of the trajectory) between jumps are exponentially distributed r.v.'s.

Definition 1.10. A *continuous time Markov chain* with finite or countable state space, S, is a family $\{X(t) : t \geq 0\}$ of random variables taking values in S, having piecewise constant sample paths with probability one, and satisfying the Markov property:

$$P\{X(t + s) = j \mid X(s) = i, X(r) = x(r); r < s\}$$
$$= P\{X(t + s) = j \mid X(s) = i\} \equiv p_{ij}(t),$$

where $i, j, x(r) \in S$.

It can be shown that the limits

(1.11) $$q_{ij} = \lim_{h \downarrow 0} \frac{1}{h} (p_{ij}(h) - p_{ij}(0))$$

exist for all i and j. The matrix

$$Q = (q_{ij} : i, j \in S)$$

describes the infinitesimal behavior of the CTMC and is called the *infinitesimal generator*. When the state space, S, is a finite set, the matrix Q generates

$$P(t) = (p_{ij}(t): i,j \in S)$$

in the sense that

$$P(t) = \exp(Qt) = \sum_{n=0}^{\infty} \frac{Q^n t^n}{n!}$$

The converse is more important in the simulation context. Given Q and p, the task is to find a CTMC with infinitesimal generator, Q, and initial probability vector, p, and to generate sample paths of the process. This can be achieved only if Q satisfies certain conditions; i.e., there are some Q's for which there exists *no* CTMC satisfying (1.11). We impose the following restrictions on Q and p:

(i) $q_{ij} \geq 0$, $i \neq j$

(ii) $\sum_{j \in S} q_{ij} = 0$, $i \in S$

(iii) $0 < q_i < +\infty$, where $q_i = -q_{ii}$

and either

(iv) $\sup\{q_i : i \in S\} < +\infty$

or

(iv') the states i satisfying $p_i > 0$ are recurrent states for the DTMC with transition matrix $R = \{r_{ij} : i,j \in S\}$, where

$$r_{ij} = \begin{matrix} q_{ij}/q_i & \text{if } j \neq i \\ 0 & \text{if } j = i \end{matrix}.$$

A semi-Markov process (SMP) is a continuous time process that represents a generalization of a CTMC. The distribution of holding times in state i need not be exponentially distributed and may depend on the next state visited, say j. A SMP is constructed in terms of a Markov renewal process (MRP). Let $S = \{S_n : n \geq 0\}$ be an increasing sequence of r.v.'s $(0 = S_0 \leq S_1 \leq ...)$ and $X = \{X_n : n \geq 0\}$ be a stochastic process with finite or countable state space, S.

Definition 1.12. The process $(X,S) = \{(X_n, S_n) : n \geq 0\}$ is a *Markov renewal process* with state space S if

$$P\{X_{n+1} = j, S_{n+1} - S_n \leq t \mid X_0, X_1, ..., X_n; S_0, S_1, ..., S_n\}$$

$$= P\{X_{n+1} = j, S_{n+1} - S_n \leq t \mid X_n\}$$

with probability one for all $n \geq 0$, $j \in S$ and $t \geq 0$.

The given data for a MRP is the kernel, Q, defined by

$$Q(i,j,t) = P\{X_{n+1} = j, S_{n+1} - S_n \leq t \mid X_n = i\}$$

for $i,j \in S$ and $t \geq 0$, along with an initial probability vector, p. As in the Markov chain case, we impose restrictions on $Q(i,j,t)$ and p so as to guarantee reasonable behavior:

(i) $Q(i,j,t)$ is a probability on $S \times [0, +\infty)$

and either

(ii) there exist $\alpha, \varepsilon > 0$ such that

$$\sup \{Q(i,j,\alpha) : i,j \in S\} < 1 - \varepsilon$$

or

(iii) the states i satisfying $p_i > 0$ are recurrent states for the DTMC with transition matrix $R = \{r_{ij} : i,j \in S\}$, where

(1.13) $$r_{ij} = \lim_{t \to \infty} Q(i,j,t).$$

Definition 1.14. Assume that the matrix Q satisfies assumptions (i), (ii), and (iii). The *semi-Markov process* $Y = \{Y(t): t \geq 0\}$ corresponding to Q and p is given by

$$Y(t) = X_n \text{ for } t \in [S_n, S_{n+1}),$$

where $\{(X_n, S_n): n \geq 0\}$ is the Markov renewal process with kernel, Q, and initial distribution, p.

A SMP visits a sequence of states in accordance with the DTMC $\{X_n: n \geq 0\}$ and spends a random amount of time in each state. Unlike the embedded jump chain of a CTMC, with positive probability $\{X_n: n \geq 0\}$ may remain in a state after one jump. Also as stated above, the holding times in state i have an arbitrary distribution (concentrated on the non-negative real line) that may depend on the next state visited.

The last type of process to be discussed here is a a renewal process (RP). Consider a system that fails after a random time and then is immediately replaced (renewed). A renewal counting process (RCP) counts the number of replacements in the interval $[0,t]$. The renewal process is the sequence of times at which the replacements occur. Let $\{T_n: n \geq 1\}$ be a sequence of positive i.i.d. r.v.'s defined on a probability space $(\Omega, \mathcal{F}, P)$. Set $S_0 = 0$ and form the partial sums $S_n = T_1 + ... + T_n$, $n \geq 1$.

Definition 1.15. The process $S = \{S_n: n \geq 0\}$ is a *renewal process*.

Definition 1.16. The process $N = \{N(t): t \geq 0\}$ is a *renewal counting process* if

$$N(t) = \sup\{n \geq 0: S_n \leq t\}.$$

A law of large numbers in essence asserts that the sample mean of a stochastic process converges either strongly (with probability one) or weakly (in probability or in distribution) to an appropriate true (population) mean. The classical laws of large numbers deal with sequences of i.i.d. r.v.'s.

Theorem 1.17. (Strong Law of Large Numbers) Let $\{X_n : n \geq 1\}$ be a sequence of independent, identically distributed random variables. Set $S_n = X_1 + X_2 + ... + X_n$, $n \geq 1$. If $E\{|X_1|\} < \infty$, then

$$\frac{S_n}{n} \to E\{X_1\} \text{ a.s.}$$

as $n \to \infty$. Furthermore, if $E\{X_1\} = +\infty$, then

$$\frac{S_n}{n} \to +\infty \text{ a.s.}$$

as $n \to \infty$.

Since this result holds with probability one, it also holds in probability, and hence in distribution.

A similar result holds for RCP's. Let $\{T_n : n \geq 1\}$ be a sequence of positive i.i.d. r.v.'s and let $S = \{S_n : n \geq 0\}$ and $N = \{N(t) : t \geq 0\}$ be the associated RP and RCP.

Theorem 1.18. (Strong Law of Large Numbers For RCP's) If $E\{T_1\} < \infty$, then

$$\frac{N(t)}{t} \to \frac{1}{E\{T_1\}} \text{ a.s.}$$

as $t \to \infty$. If $E\{T_1\} = +\infty$, then

$$\frac{N(t)}{t} \to 0 \text{ a.s.}$$

as $t \to \infty$.

Corresponding results hold for DTMC's, CTMC's, and SMP's.

Definition 1.19. The stochastic vector π is said to be a *stationary distribution* of a discrete time Markov chain with transition matrix P if π is a solution of $\pi P = \pi$. The stochastic vector π is said to be a *stationary distribution* of a continuous time Markov chain with infinitesimal generator Q if π is a solution of $\pi Q = 0$.

Theorem 1.20. (Strong Law of Large Numbers for DTMC's) Let $\{X_n : n \geq 0\}$ be an irreducible, positive recurrent discrete time Markov chain with stationary distribution, π. If f is a real-valued function such that

$$\sum_{i \in S} \pi_i |f(i)| < \infty,$$

then

$$\frac{1}{n} \sum_{i=0}^{n-1} f(X_i) \to \sum_{i \in S} \pi_i f(i) \quad \text{a.s.}$$

for any initial distribution, p.

Theorem 1.21. (Strong Law of Large Numbers for CTMC's) Let $\{X(t) : t \geq 0\}$ be an irreducible, positive recurrent continuous time Markov chain with stationary distribution, π. If f is a real-valued function such that

$$\sum_{i \in S} \pi_i |f(i)| < \infty,$$

then

$$\frac{1}{t} \int_0^t f(X(u)) du \to \sum_{i \in S} \pi_i f(i) \quad \text{a.s.}$$

for any initial distribution p.

For an SMP, set

$$F_{ij}(t) = \frac{Q(i,j,t)}{r_{ij}}$$

for $r_{ij} > 0$ and let m_{ij} be the mean of the distribution F_{ij}. (Set $m_{ij} = 0$ for $r_{ij} = 0$.)

Theorem 1.22. (Strong Law of Large Numbers for SMP's) Let $\{Y(t): t \geq 0\}$ a SMP for which the matrix $R = \{r_{ij} : i, j \in S\}$ is irreducible and positive recurrent with stationary distribution, γ. Assume that

$$\sum_{i,j \in S} \gamma_i \, r_{ij} \, m_{ij} < \infty$$

and let

$$m_i = \sum_{j \in S} r_{ij} \, m_{ij}$$

and

$$\pi_j = \frac{\gamma_j \, m_j}{\sum_{i \in S} \gamma_i \, m_i}.$$

If f is a real-valued function such that

$$\sum_{i \in S} \pi_i \, |f(i)| < \infty,$$

then

$$\frac{1}{t} \int_0^t f(Y(u)) \, du \to \sum_{i \in S} \pi_i \, f(i) \quad \text{a.s.}$$

for any initial distribution, p.

Central limit theorems are in a sense second order results about the random fluctuations (measured in an appropriate scale) of the sample mean about the population mean. Such theorems, however, are always weak convergence rather than strong convergence results. The classical c.l.t. deals with an i.i.d. sequence $\{X_n : n \geq 1\}$ of r.v.'s having finite mean, μ, and finite variance, σ^2. Again set

$$S_n = X_1 + X_2 + \ldots + X_n,$$

$n \geq 1$. Denote by $N(0,1)$ a standardized (mean zero, variance one) normal random variable.

Theorem 1.23. (Central Limit Theorem) Let $\{X_n : n \geq 1\}$ be a sequence of independent, identically distributed random variables. If $\mu = E\{X_1\} < \infty$ and $\sigma^2 = \text{var}(X_1) < \infty$, then

(1.24) $$\frac{(S_n - n\mu)}{n^{1/2}} \Rightarrow \sigma N(0,1)$$

as $n \to \infty$.

Theorem 1.25 provides a c.l.t. for a random number of summands. Let $\{\nu_n : n \geq 1\}$ be an i.i.d. sequence of be a sequence of positive, integer-valued r.v.'s with finite mean, μ, and finite variance, σ^2. The sequences $\{\nu_n : n \geq 1\}$ and $\{X_n : n \geq 1\}$ need not be independent.

Theorem 1.25. Let $0 < c < \infty$. If $\nu_n / n \Rightarrow c$, then

(1.26) $$\frac{(S_{\nu_n} - \nu_n \mu)}{\sigma \nu_n^{1/2}} \Rightarrow N(0,1)$$

and

(1.27) $$\frac{(S_{\nu_n} - \nu_n \mu)}{\sigma(cn)^{1/2}} \Rightarrow N(0,1)$$

as $n \to \infty$.

Observe that (1.26) is obtained simply by replacing n in (1.24) by ν_n. Also observe that (1.27) is obtained from (1.26) by an application of Lemma 1.7 and the continuous mapping theorem.

Theorem 1.28 is known as the convergence of types theorem. Let $\{Z_n : n \geq 0\}$ be an arbitrary sequence of r.v.'s and Y and Z be two non-degenerate r.v.'s.

Theorem 1.28. (Convergence of Types) Let $Z_n \Rightarrow X$ and $a_n Z_n + b_n \Rightarrow Y$, where $a_n > 0$ and b_n are constants. Then there exist constants a and b such that $a_n \to a$, $b_n \to b$, and Y has the same distribution as $aX + b$.

In the simulation context, the principal application of this result is to show that the variance constants in two c.l.t.'s are the same. Suppose that $\{U_n : n \geq 1\}$ is a sequence of r.v.'s and that we have shown that

$$\frac{U_n}{n^{1/2}c} \Rightarrow N(0,1)$$

and

$$\frac{U_n}{n^{1/2}c'} \Rightarrow N(0,1)$$

as $n \to \infty$, where $c, c' > 0$. Take the r.v.'s X and Y to be $N(0,1)$ and set

$$Z_n = U_n / (n^{1/2} c),$$

$a_n = a = c/c' > 0$, and $b_n = 0$. Then by the convergence of types theorem, Y and $(c/c')X$ have the same distribution. Therefore, $c = c'$. If the two c.l.t.'s are indexed by a continuous parameter, the same argument goes through.

Appendix 2

Convergence of Passage Times

Label the jobs from 1 to N and for $t \geq 0$ let $Z(t)$ be the vector in (1.1) of Section 3.1 that defines the job stack at time t. Define the vectors

$$Y^i(t) = (Z(t), N^i(t)),$$

where $N^i(t)$ is the position of the job labelled i in the job stack at time t, $i = 1,2,...,N$. Assume that for some state z^*, the set D of all states of the job stack process accessible from z^* and the set $G = \{(z,n): z \in D, 1 \leq n \leq N\}$ are irreducible in the sense that all pairs of states communicate. Also assume that the GSMP's associated with the processes $Y^i = \{Y^i(t): t \geq 0\}$ are regenerative processes defined on a common underlying probability triple, $(\Omega, \mathscr{F}, P)$, say.

Observe that if the marked job is the job labelled i, then the process Y^i coincides with the augmented job stack process X, except possibly for the initial condition at $t = 0$. Define for each job two sequences of times, the starts and terminations of the successive passage times for the job. For the job labelled i, denote these times by $\{S_j^i : j \geq 0\}$ and $\{T_j^i : j \geq 1\}$. The definition of these times in terms of the process Y^i is completely analogous to what was done in Section 4.1 in terms of the process X. Then the jth passage time for the job labelled i is $P^i_j = T^i_j - S^i_{j-1}$, $j \geq 1$. For the job labelled i, let X^i_j be the state of the GSMP Y^i when the $(j+1)$st passage time starts for job i: $X^i_j = Y^i(S^i_j)$. All of the discrete time processes $\{(X^i_j, S^i_j) : j \geq 0\}$, are defined on $(\Omega, \mathcal{F}, P)$.

Next we introduce a new sequence of passage times, $\{P'_j : j \geq 1\}$, also defined on $(\Omega, \mathcal{F}, P)$. This is the sequence of passage times irrespective of job identity, enumerated in order of start time. For each j, P'_j is a random member of the set $\{P^i_l : 1 \leq l \leq j; 1 \leq i \leq N\}$; this means that $P'_j = P^{l(j)}_{k(j)}$, where $l(j)$ and $k(j)$ are random variables.

The principal result of this Appendix is to show that all the sequences $\{P'_j : j \geq 1\}$ and $\{P^i_j : j \geq 1\}$ converge in distribution to a common random variable P.

Proposition 2.1. For $i = 1, 2, \ldots, N$, $P^i_j \Rightarrow P$ as $j \to \infty$.

Proof. Since the N jobs are identical with respect to their service requirements and routing probabilities, the probability mass functions and clock setting distributions governing the GSMP's associated with the processes Y^i are the same. This implies that the processes $\{(X^i_j, S^i_j) : j \geq 0\}$ have the same finite dimensional distributions. In fact, for any particular job the only difference is that (with possibly one exception) the job does not start a passage time at $t = 0$. However, this difference does not alter limiting results; the job labelled i starts a passage time with probability one (since the GSMP associated with Y^i returns to every state infinitely often with probability one) and once this occurs, the situation is the same. Note in particular that $S^i_j \to +\infty$ with probability one for all i; thus, there is always a next passage time for every job. This being so, we have $P^i_j \Rightarrow P$ as $j \to \infty$.

Next we show that $P'_j \Rightarrow P$. Since $P^i_j \Rightarrow P$ for all i, we can use the Skorohod representation theorem to assert the existence of a probability space $(\tilde{\Omega}, \tilde{\mathcal{F}}, \tilde{P})$, and random variables $\tilde{P}^i_j$ ($j \geq 1$, $1 \leq i \leq N$) and $\tilde{P}$ defined on that space such that (i) $\tilde{P}^i_j$ and $\tilde{P}$ have

212 Convergence of Pasage Times

the same distribution as P_j^i and P, respectively, and (ii) $\tilde{P}_j^i \to \tilde{P}$ with probability one as $j \to \infty$ for all i. These representatives $\tilde{P}_j^i$ also provide representatives that we call $\tilde{P}_j'$ for the P_j'.

Putting aside the null sets of Ω on which the above convergence statements do not hold, we examine the numerical sequence $\{\tilde{P}_j'(\omega): j \geq 1\}$ for one of the remaining $\omega \in \Omega$. We use the following criterion for convergence of a numerical sequence $\{x_j : j \geq 1\} : x_j \to x$ as $j \to \infty$ if and only if for each subsequence $\{x_{j'}\}$ there exists a further subsequence $\{x_{j''}\}$ that converges to x. Select a subsequence $\{\tilde{P}_j'(\omega)\}$. This subsequence must contain a further subsequence $\{\tilde{P}_k''(\omega)\}$ that is identical to a subsequence of one of the sequences $\{\tilde{P}_j^i(\omega): j \geq 1\}$, say for $i = i_0$. This follows from the fact that there are only a finite number of jobs. But this subsequence $\{\tilde{P}_j^{i_0}\}$ converges with probability one to $\tilde{P}(\omega)$ since the full sequence does. Thus $\tilde{P}_j' \to \tilde{P}$ a.s. and therefore $P_j' \Rightarrow P$. □

Bibliography

Asmussen, S. (1986). *Applied Probability and Queueing.* John Wiley. New York, New York.

Billingsley, P. (1979). *Probability and Measure.* John Wiley. New York, New York.

Breiman, L. (1968). *Probability.* Addison-Wesley. Reading, Massachusetts.

Chung, K. L. (1974). *A Course in Probability Theory.* Second Edition. Academic Press. New York, New York.

Çinlar, E. (1975). *Introduction to Stochastic Processes.* Prentice-Hall. Englewood Cliffs, New Jersey.

Crane, M. A. and Iglehart, D. L. (1975). Simulating stable stochastic systems: III, Regenerative processes and discrete event simulation. *Operations Res.* **23**, 33-45.

Crane, M. A. and Iglehart, D. L. (1975). Simulating stable stochastic systems, IV: Approximation techniques. *Management Sci.* **21**, 1215-1225.

Crane, M. A. and Lemoine, A. J. (1977). *An Introduction to the Regenerative Method for Simulation Analysis.* Lecture Notes in Control and Information Sciences, Vol. 5. Springer-Verlag. Berlin, Heidelberg, New York.

Doob, J. L. (1953). *Stochastic Processes.* John Wiley and Sons, Inc. New York, New York.

Fossett, L. (1979). Simulating generalized semi-Markov processes. Ph. D. Dissertation. Department of Operations Research. Stanford University. Stanford, California.

Glynn, P. W. (1982). Regenerative aspects of the steady-state simulation problem for Markov chains. Technical Report No. 61. Department of Operations Research. Stanford University. Stanford, California.

Glynn, P. W. (1982). Regenerative simulation of Harris recurrent Markov chains. Technical Report No. 62. Department of Operations Research. Stanford University. Stanford, California.

Glynn, P. W. (1985). Regenerative structure of Markov chains simulated via common random numbers. *Oper. Res. Lett.* **4**, 49-53.

Gunther, F. L. and Wolff, R. W. (1980). The almost regenerative method for stochastic system simulations. *Operations Res.* **28**, 375-386.

Haas, P. J. (1985). Recurrence and regeneration in non-Markovian simulations. Ph. D. Dissertation. Department of Operations Research. Stanford University. Stanford, California.

Haas, P. J. and Shedler, G. S. (1985). Regenerative simulation methods for local area computer networks. *IBM J. Res. Develop.* **29**, 194-205.

Haas, P. J. and Shedler, G. S. (1985). Recurrence and regeneration in non-Markovian networks of queues. IBM Research Report RJ4671. San Jose, California.

Heidelberger, P. (1979). A variance reduction technique that increases regeneration frequency. *Current Issues in Computer Simulation*. N. Adam and A. Dogramaci, eds. Academic Press. New York.

Heidelberger, P. (1980). Variance reduction techniques for the simulation of Markov processes, I: Multiple estimates. *IBM J. Res. Develop.* **24**, 570-581.

Heidelberger, P. and Iglehart, D. L. (1979). Comparing stochastic systems using regenerative simulation with common random numbers. *Adv. Appl. Probability* **11**, 804-819.

Heidelberger, P. and Lewis, P. A. W. (1981). Regression-adjusted estimates for regenerative simulations, with graphics. *Comm. Assoc. Comput. Mach.* **24**, 260-273.

Hordijk, A., Iglehart, D. L. and Schassberger, R. (1976). Discrete time methods for simulating continuous time Markov chains. *Adv. Appl. Probability* **8**, 772-788.

Iglehart, D. L. (1975). Simulating stable stochastic systems, V: Comparison of ratio estimators. *Naval Res. Logist. Quart.* **22**, 553-565.

Iglehart, D. L. (1976). Simulating stable stochastic systems, VI: Quantile estimation. *J. Assoc. Comput. Mach.* **23**, 347-360.

Iglehart, D. L. (1977). Simulating stable stochastic systems, VII: Selecting the best system. In *Algorithmic Methods in Probability*. M. F. Neuts (ed.), 37-50. North-Holland. Amsterdam.

Iglehart, D. L. (1978). The regenerative method for simulation analysis. In *Current Trends in Programming Methodology Vol. III: Software Engineering*. K. M. Chandy and R. T. Yeh (eds.), 52-71. Prentice-Hall. Englewood Cliffs, New Jersey.

Iglehart, D. L. and Lewis, P. A. W. (1978). Regenerative simulation with internal controls. *J. Assoc. Comput. Mach.* **26**, 271-282.

Iglehart, D. L. and Shedler, G. S. (1980). *Regenerative Simulation of Response Times in Networks of Queues*. Lecture Notes in Control and Information Sciences, Vol. 26. Springer-Verlag. Berlin, Heidelberg, New York.

Iglehart, D. L. and Shedler, G. S. (1981). Regenerative simulation of passage times in networks of queues: statistical efficiency. *Acta Informatica* **15**, 347-363.

Iglehart, D. L. and Shedler, G. S. (1983). Simulation for passage times in non-Markovian networks of queues. *Proc. IFIP Workshop on Stochastic Programming: Algorithms and Applications*. Gargnano, Italy.

Iglehart, D. L. and Shedler, G. S. (1983). Statistical efficiency of regenerative simulation methods for networks of queues. *Adv. Appl. Probability* **15**, 183-197.

Iglehart, D. L. and Shedler, G. S. (1983). Simulation of non-Markovian systems. *IBM J. Res. Develop.* **27**, 427-480.

Iglehart, D. L. and Shedler, G. S. (1984). Simulation output analysis for local area computer networks. *Acta Informatica* **21**, 321-338.

Iglehart, D. L. and Stone, M. L. (1983). Regenerative simulation for estimating extreme values. *Operations Res.* **31**, 1109-1144.

Karlin, S. and Taylor, H. M. (1975). *A First Course in Stochastic Processes*. Academic Press. New York, New York.

König, D., Matthes, K., and Nawrotzki, K. (1967). *Verallgemeinerungen der Erlangschen und Engsetschen Formeln*. Akademie-Verlag. Berlin.

König, D., Matthes, K., and Nawrotzki, K. (1974). Unempfindlichkeitseigenschaften von Bedienungsprozessen. Appendix to Gnedenko, B. V. and Kovalenko, I. N., *Introduction to Queueing Theory*, German edition.

Lavenberg, S. S. and Sauer, C. H. (1977). Sequential stopping rules for the regenerative method of simulation. *IBM J. Res. Develop.* **21**, 545-558.

Lavenberg, S. S., Moeller, T. L. and Sauer, C. H. (1979). Concomitant control variables applied to the regenerative simulation of queueing systems. *Operations Res.* **27**, 134-160.

Lemoine, A. J. (1977). The regenerative method and the workload process in general single-server queueing systems. Technical Report No. 86-26. Control Analysis Corporation. Palo Alto, California.

Matthes, K. (1962). Zur Theorie der Bedienungsprozesse. *Trans. 3rd Prague Conference on Information Theory and Statistical Decision Functions.* Prague.

Miller, D. R. (1972). Existence of limits in regenerative processes. *Ann. Math. Statist.* **43**, 1275-1282.

Shedler, G. S. and Slutz, D. R. (1981). Irreducibility in closed multiclass networks of queues with priorities: passage times of a marked job. *Performance Evaluation* **1**, 334-343.

Shedler, G. S. and Southard, J. (1982). Simulation for passage times in closed, multiclass networks of queues with unrestricted priorities. *Performance Evaluation* **2**, 257-267.

Shedler, G. S. and Southard, J. (1982). Regenerative simulation of networks of queues with general service times: passage through subnetworks. *IBM J. Res. Develop.* **26**, 625-633.

Smith, W. L. (1958). Renewal theory and its ramifications. *J. Royal Statist. Soc. Ser. B* **20**, 243-302.

Whitt, W. (1980). Continuity of generalized semi-Markov processes. *Math. Oper. Res.* **5**, 494-501.

Symbol Index

A_1, A_2	subsets defining starts of passage time for marked job	77
$A_1^{(v)}, A_2^{(v)}$	subsets defining starts of passage times for type v marked job	116
B_1, B_2	subsets defining starts of passage time for marked job	77
$B_1^{(v)}, B_2^{(v)}$	subsets defining terminations of passage times for type v marked job	116
c	number of job classes	59
C	set of (center, class) pairs	59
C_n	vector of clock readings at time ζ_n	3
$C_{j_1}^{(i)}(t)$	number of jobs of class $j_1(i)$ in queue at center i at time t	60
$C(s)$	vector of possible clock readings in state s	4
$d(z;z')$	distance from state z to z'	68
D	set of recurrent states of job stack process	139
D^*	state space of job stack process	61
e^*	fixed trigger event	3
e	event of GSMP	
E	event set	3
$E(s)$	set of events scheduled in state s	3
f	real valued (measurable) function	59
$F(\cdot;s',e',s,e^*)$	clock setting distribution	3

Symbol Index 219

G	set of recurrent states of augmented job stack process	151
G^*	state space of augmented job stack process	77
G^0	state space of fully augmented job stack process	99
$j_{k(i)}(i)$	lowest priority job class at center i	60
$L(t)$	last state of augmented job stack process before jumping to $X(t)$	117
$M(t)$	last state of job stack process before jumping to $Z(t)$	143
N	number of jobs	59
$N(s';s,e^*)$	set of new events	4
$N(t)$	position of marked job at time t	77
$N^i(t)$	position of job i at time t	98
$O(s';s,e^*)$	set of old events	5
$p_{ij,kl}$	job routing probability	59
$p(s';s,e)$	state transition probability	3
P	limiting passage time for marked job	78
P_j	jth passage time for marked job	78
$P^{(\nu)}$	limiting passage time for type ν job	118
$\mathbf{P}$	job routing matrix	59
$\mathbf{P}^\nu$	job routing matrix for type ν jobs	110
q_{ij}	infinitesimal parameter for CTMC	62
$Q_i(t)$	number of jobs at center i at time t	60
$\mathbf{Q}$	infinitesimal generator for CTMC	62
$r(f)$	limiting expectation for function f	28
$\hat{r}(n)$	point estimate for $r(f)$ based on n cycles	29
$R^{(\nu)}$	limiting response time for type ν job	118

Symbol Index

Symbol	Description	Page
s	number of service centers	59
$s(n)$	point estimate for σ based on n cycles	30
S	state space of GSMP	3
$S_i(t)$	class of job in service at center i at time t	60
S_{j-1}	start time for jth passage time for marked job	77
$S_{j-1}^{(v)}$	start time for jth passage time for type v marked job	117
$\mathscr{S}$	set of distribution functions with absolutely continuous component	25
$\sigma, \sigma(f)$	variance constant for function f	28
T_j	termination time for jth passage time for marked job	77
$T_j^{(v)}$	termination time for jth passage time for type v marked job	117
τ_k	length of the kth cycle	24
$V(t)$	(last state, current state) of augmented job stack process	117
x_0	single state of augmented job stack process	152
x^*	target state of augmented job stack process	84
X	limiting distribution for augmented job stack process	88
$X(t)$	augmented job stack at time t	77
$X^0(t)$	fully augmented job stack at time t	77
$Y_k(f)$	integral over kth cycle for function f	26
$Y(t)$	(last state, current state) of job stack process	143
z_0	single state of job stack process	141
z^*	target state of job stack process	67
Z	limiting distribution for job stack process	72
$Z(t)$	job stack at time t	60
ζ_n	nth state transition time	3

Subject Index

Central limit theorem
    for i.i.d random variables, 207
    for random sums, 208
Communicating states
    of job stack process, 139
    of augmented job stack process, 150
Confidence interval
    in regenerative simulation, 31
Continuous mapping theorem, 198
Convergence
    in distribution, 198
    in probability, 197
    with probability one, 196
    of types theorem, 198
Converging together lemma, 208
Cyclic queues, 47, 51, 57, 60, 179
    preemptive, 16
    two-server, 9, 23
    with feedback, 7, 21, 78
    with two job types, 125
    with nonpreemptive priority, 159, 183
    with preemptive priority, 60

Data base management system model, 71, 75, 80, 86
Distribution function
    aperiodic, 25
    new better than used, 57

Generalized semi-Markov process
 definition of, 6
 general state space Markov chain of, 5
 irreducible, 7
 limit theorem for, 7
 recurrence theorem for, 141
 regenerative, 50, 53
Geometric trials
 lemma, 52
 recurrence criterion, 53

Job stack, 60
Job stack process, 60
 augmented, 77
 embedded jump chain of, 67
 fully augmented, 177
 irreducibility of, 67
 irreducibility of augmented, 84
 simulation of Markovian, 62
 target state of, 66

Labelled jobs method
 central limit theorem for, 102
 for Markovian networks, 102
 for non-Markovian networks, 182
 statistical efficiency of, 188

Marked job method
 for Markovian networks, 90
 for non-Markovian networks, 157
 statistical efficiency of, 188
Markov arrival process, 165
Markov chain
 continuous time, 199
 discrete time, 200
 irreducible, 200
 infinitesimal generator for, 201
 recurrent state of, 200
 stationary distribution for, 205

Markov renewal process, 202

Networks of queues
    closed, 59
    finite capacity open, 164
    with multiple job types, 110

Passage times
    bivariate central limit theorem for, 121
    convergence of, 211
    ratio formula for, 88, 101
    specification of, 77
Phantom server, 169

Queue
    multi-server, 19
    single-server with scheduled arrivals, 24

Recurrent states
    of job stack process, 75
    of augmented job stack process, 92
    irreducible, closed sets of, 67, 84
Regeneration points
    definition of, 21
    sequences of, 35
Regenerative process
    definition of, 21
    limit theorem for, 26
    ratio formula for, 27
    standard deviation constant for, 31
Regenerative method
    central limit theorem for, 30, 32
    confidence interval in, 31
    for Markovian networks, 74
    for non-Markovian networks, 146
    point estimate in, 29
    standard, 32
    statistical efficiency of, 47

Renewal counting process
    definition of, 204
    strong law for, 205
Renewal process, 204
Routing matrix, 59

Semi-Markov process
    definition of, 203
    strong law for, 206
Single state
    of augmented job stack process, 152
    of job stack process, 141
    of job stack process for passage time, 175
Stopping time, 20
Strong law of large numbers, 204
    for discrete time Markov chain, 205
    for continuous time Markov chain, 206
    for renewal counting process, 205
    for semi-Markov process, 206
System overhead model, 12, 22, 32

Tandem queues, 171

RAYMOND H. FOGLER LIBRARY
DATE DUE

BOOKS ARE SUBJECT TO RECALL AFTER TWO WEEKS

APR 2 3 1987